SCIENTISTS
AND THEIR DISCOVERIES

Christine Hatt

Evans

Published in 2007 Evans Brothers Limited
2A Portman Mansions
Chiltern Street
London, W1U 6NR

Printed in Dubai by Oriental Press

British Library Cataloguing in Publication Data

Hatt, Christine
   Scientists and their discoveries. - (History in writing)
   1. Scientists - Biography - Juvenile literature
   2. Discoveries in science - Juvenile literature
   I. Title
   509.2'2

ISBN 023753195X
ISBN 13- 9780237531959

Design – Neil Sayer
Editorial – Nicola Barber
Illustrations – Richard Morris
Consultant – Peter R. Ellis (science teacher and writer, and past chairman of the British Society for the History of Science, Education Section)
Production – Jenny Mulvanny

Title page picture: Albert Einstein
Border: Microscope and genes

To find out more about the scientists mentioned in this book try looking at these websites:

http://dir.yahoo.com/Science/People/

http://www.pbs.org/wgbh/aso/databank/bioindex.html
(People and Discoveries)

http://es.rice.edu/ES/humsoc/Galileo/Catalog/ catalog.html
(catalogue of 16th- and 17th-century scientists)

http://www.webelements.com
(for information about the Periodic Table)

VISIT OUR WEBSITE
Evans
www.evansbooks.co.uk

ACKNOWLEDGEMENTS

For permission to reproduce pictorial material, the author and publishers gratefully acknowledge the following:

Cover: (top left: Marie Curie) Science Photo Library (bottom left: human genes) Science Photo Library (centre: Nobel medal) Science & Society Picture Library (top right: Faraday lecture) Science & Society Picture Library (bottom right: Roger Penrose) Science & Society Picture Library **title page** (Albert Einstein) Science and Society Picture Library **page 8** Science & Society Picture Library **page 9** (top) Science Photo Library (bottom) Science Photo Library **page 10** Science Photo Library **page 11** Science Photo Library **page 12** Science Photo Library **page 13** Topham Picturepoint **page 14** Science Photo Library **page 15** (top) Science & Society Picture Library (bottom) Science & Society Picture Library **page 16** Science Photo Library **page 17** (top) Science Photo Library (bottom) Science Photo Library **page 18** Science Photo Library **page 19** (top) Science & Society Picture Library (bottom) Science Photo **page 20** Science Photo Library **page 21** (left) Science & Society Picture Library (right) Science Photo Library **page 22** (top) Science & Society Picture Library (bottom) Science Photo Library **page 23** Mary Evans Picture Library **page 24** Science Photo Library **page 25** Corbis **page 26** Science & Society Picture Library **page 27** Science Photo Library **page 28** Science & Society Picture Library **page 29** (left) Science Photo Library (right) Science Photo Library **page 30** Science Photo Library **page 31** (top) Science Photo Library (middle) Science & Society Picture Library **page 32** Science Photo Library **page 33** (left) Science Photo Library (right) Science & Society Picture Library **page 34** Science & Society Picture Library **page 35** Science & Society Picture Library **page 36** (left) Science Photo Library (right) Science Photo Library **page 38** (left) Science Photo Library (right) Mary Evans Picture Library **page 39** Oxford Scientific Films **page 40** Science Photo Library **page 42** Geology Department, Edinburgh University **page 43** Bruce Coleman Collection **page 45** (left) Photri Microstock (right) Science Photo Library **page 46** Science & Society Picture Library **page 48** Science & Society Picture Library **page 49** (left) Science Photo Library (right) Bruce Coleman Collection **page 50** Science Photo Library **page 51** (top) Mary Evans Picture Library (bottom) Science & Society Picture Library **page 52** (left) Science & Society Picture Library (right) Science Photo Library **page 53** Photri Microstock **page 54** (left) Science & Society Picture Library (right) Science Photo Library **page 55** (top) Science & Society Picture Library (bottom) Science & Society Picture Library **page 56** Science Photo Library **page 57** Science Photo Library **page 58** (top) Science Photo Library (bottom) Science Photo Library **page 59** Topham Picturepoint

For permission to reproduce copyright material for the documents, the author and publisher gratefully acknowledge the following:

page 9 From Cecilia Payne-Gaposchkin: an autobiography and other recollections, edited by Katherine Haramundais, published in 1996 by Cambridge University Press, © Cambridge University Press,1984,1996 **page 11** (top) From The Great Copernicus Chase and other adventures in astronomical history by Owen Gingerich, published in 1992 by the Sky Publishing Corporation and Cambridge University Press, © Sky Publishing Corporation 1992 **page 11** (bottom) From the Selected Papers of Edwin Hubble, from the Edwin Hubble Collection, Huntington Library, San Marino, California **page 12** From Lemaître, Big Bang and the Quantum Universe, by Michael Heller, published in 1996 by Pachart Publishing House, © 1996 by the Pachart Foundation **page 14** From The Creation of the Universe, by George Gamow, published by the Viking Press, New York, 1952, © Cambridge University Press 1965 **page 15** From Mr Tompkins in Paperback, by George Gamow, published in 1965 by Cambridge University Press, © George Gamow 1952 **page 17** From A Brief History of Time, by Stephen Hawking, published in 1996 by Bantam Press, a division of Transworld Publishers Ltd, © Stephen Hawking 1988, 1996 **page 19** From A New System of Chemical Philosophy Part 1, by John Dalton, published in 1803 by R. Bickerstaff, London **pages 23, 24 and 25** From Madame Curie, by Eve Curie, published in 1938 by William Heinemann Ltd **page 27** (middle and bottom) From Dorothy Hodgkin: A Life, by Georgina Ferry, published in Great Britain by Granta Books 1998, © Georgina Ferry 1998 **page 29** From Michael Faraday and the Royal Institution, by John Meurig-Thomas, published in 1991 by Adam Hilger, © IOP Publishing Ltd 1991 **page 31** (middle, bottom) From The Demon in the Aether, The Story of James Clerk Maxwell, by Martin Goldman, published in 1983 by Paul Harris Publishing , © 1983 Martin Goldman **page 35** From Introducing Einstein, by Joseph Schwartz and Michael McGuiness, edited by Richard Appignanesi. Published in 1999 by Icon Books Ltd, © 1979 Joseph Schwartz (text), © 1979 Michael McGuiness (illustrations) **page 39** (middle, bottom) From Life, Letters and Journals of Sir Charles Lyell, Bart, edited by Mrs Lyell, published in 1881 by John Murray, republished 1970 by Gregg International Publishers Ltd **page 41** From The Origin of Continents and Oceans, by Alfred Wegener, published in 1924 by Methuen & Co Ltd, © 1966 Dover Publications **page 42 and 43** From The Collected Papers of Lord Rutherford of Nelson, published in 1963 by George Allen & Unwin Ltd, © 1963 George Allen & Unwin Ltd **page 45** From Studies in Earth and Space Sciences, edited by R. Shagam, R.B. Hargraves, W.G. Morgan, F.B. Van Houten, C.A. Burk, H.D. Holland, L.C. Hollister, published by The Geographical Society of America, Inc, © The Geographical Society of America 1972 **page 47 and 49** From The Origin of Species By Means of Natural Selection, by Charles Darwin, first published by John Murray in 1859 **page 51** From Experiments in Plant Hybridisation, by Gregor Mendel, edited by J.H. Bennett, published by Oliver & Boyd 1965, © 1965 Oliver & Boyd **page 53** From Louis Pasteur, by Beverley Birch, published in 1990 by Exley Publications Ltd, © Exley Publications 1990 **page 55** From Penicillin: Its Practical Application, by Professor Sir Alexander Fleming, published in 1946 by Butterworth & Co **page 57** From What Mad Pursuit, by Francis Crick, published in 1998 by Basic Books Inc, Publishers, © 1988 by Francis Crick **page 59** (left) From The Harrap Science Encyclopedia (right) From the McGraw-Hill Encyclopedia of World Biography, © 1973 by McGraw-Hill, Inc

While every effort has been made to secure permission to use copyright material, Evans Brothers apologise for any errors or omissions in the above list and would be grateful for notification of any corrections to be included in subsequent editions.

CONTENTS

LOOKING AT DOCUMENTS

This book is about some of the great scientists whose discoveries transformed the world in the 19th and 20th centuries. It looks at their personal backgrounds, at the choices they made and obstacles they overcame to reach their goals. It examines their ideas and achievements, and how they affected society as a whole. Above all, it highlights their work, the mixture of experiment, collaboration and brilliant individual insight that has characterised scientific advance down the ages.

The earliest scientists investigated any aspect of the natural world that aroused their interest, from animals to planets. However, by the 19th century, scientific knowledge was growing at such a rate that people specialised in particular areas of research. *Scientists and their Discoveries* reflects this modern approach by dividing its content into five chapters, each dedicated to a different type of scientific expert. They are astronomers, chemists, physicists, geologists and biologists.

During the period covered by this book, each of these groups made major advances. Astronomers such as George Gamow developed theories about how the universe began, while chemists such as John Dalton made significant progress towards understanding the nature of matter. Some physicists delved further into this mystery, while others revealed the secrets of forces such as electricity and magnetism. Albert Einstein, the greatest physicist of the era, gained a revolutionary new insight into the natural laws that govern everything in the universe.

Geologists of this period made discoveries that were just as revolutionary. By examining the Earth itself, Charles Lyell and others overthrew ancient ideas about how landscapes and continents formed. Biologists, too, made a substantial mark. Charles Darwin, for example, developed his far-reaching theory of evolution, while in the 1950s Francis Crick and James Watson discovered the structure of DNA, the basic chemical building block of life.

To bring the stories of these people to life, *Scientists and their Discoveries* uses a wide range of documents. They include handwritten notebooks, letters of encouragement and congratulation between experts working in the same field, famous books that announced scientific breakthroughs to the general public, papers and magazines designed for specialists, and biographies.

To make the documents easy to read and understand, we have printed them all in the same type. However, you will find photographs of some of the original documents, often including illustrations, alongside several of the extracts. Difficult or technical words and phrases are explained in the captions around the documents.

On these pages are a few extracts from the documents used in this book. They have been selected to give you an idea of the great variety of documents included, and to explain how and why some of the documents were written.

Books of popular science, in which experts write about complex scientific ideas in simple terms, are often of great interest. In *Mr Tompkins in Wonderland*, astronomer and physicist George Gamow brings arguments about the universe to life by using drawings and poetry (see page 15).

What does this mean? Some words or phrases in the documents are difficult to understand. Captions alongside the documents give explanations of highlighted areas of text. You can find out what this line means on page 15.

My telescope
Has dashed your hope;
   Your tenets are refuted.
Let me be terse:
Our universe
   Grows daily more diluted!'

Scientists often kept handwritten notebooks in which they recorded the details of their experiments. These reveal the basic data on which their theories were built. In this notebook (right), the 19th-century chemist John Dalton wrote down the results of his colour blindness tests (see page 19).

Scientists also use notebooks to jot down ideas for their theories. Ernest Rutherford drew this diagram of atomic structure (left) in 1910 or 1911 (see page 33). It was based on the results of experiments he had completed in his laboratory at Manchester University.

Biographies often contain a great deal of information about scientists' personal as well as their working lives. For example, Eve Curie's biography of her mother, the great Polish-French chemist Marie Curie, shows her to be a doting parent as well as a committed scientist (see page 23).

Irène has cut her seventh tooth, on the lower left...

The various reasons we have just enumerated lead us to believe that the new radioactive substance contains a new element to which we propose to give the name of RADIUM...

Books in which great scientists reveal and explain their new ideas are extremely important documents. Among the most influential was *The Origin of Species* by Charles Darwin (see page 49). In 1859 he used this work to announce his theory of evolution to the world.

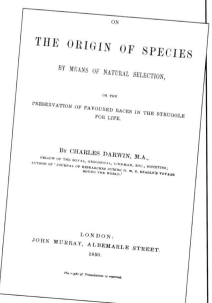

ON

THE ORIGIN OF SPECIES

BY MEANS OF NATURAL SELECTION,

OR THE

PRESERVATION OF FAVOURED RACES IN THE STRUGGLE FOR LIFE.

By CHARLES DARWIN, M.A.,

FELLOW OF THE ROYAL, GEOLOGICAL, LINNÆAN, ETC., SOCIETIES;
AUTHOR OF 'JOURNAL OF RESEARCHES DURING H. M. S. BEAGLE'S VOYAGE
ROUND THE WORLD.'

LONDON:
JOHN MURRAY, ALBEMARLE STREET.
1859.

The right of Translation is reserved.

ASTRONOMERS
HENRY RUSSELL
AND THE STUDY OF THE STARS

Until the 1800s, astronomers could do little more than survey the night skies and plot the movements of the stars. But the birth of astrophysics, the study of what stars are made of and how they evolve, opened up many possibilities. One of the first great astrophysicists was an American, Henry Norris Russell (1877-1957).

Astrophysics was made possible by spectroscopy. Scientists in the 19th century knew that white light was made up of colours and that it could be split to show this spectrum. In 1814, German physicist Josef von Fraunhofer split light from the Sun and found that there were dark lines in the colour bands. Later research showed that these lines were caused by different elements in the Sun absorbing some of the light. The pattern of the lines gave scientists information about which elements were present in the Sun.

It became clear that by analysing the spectra of other stars, scientists could learn what they were made of, too.

By the early 20th century, astronomers at Harvard University in the USA were busy classifying stars according to their spectra. Their work came to the attention of Henry Russell, Professor of Astronomy at Princeton University. He was especially interested in the life cycles of stars, and thought that the new classifications might help him to find out more. In 1913, Russell drew a graph using the Harvard data. He worked out each star's position by plotting its spectrum classification (its colour) against its luminosity (true brightness). As a star's colour reveals its temperature – blue stars are hottest and red coolest – Russell was also plotting temperature against brightness. He found that most stars fell in a diagonal from top left to bottom right – the main sequence.

There were exceptions. Some stars appeared on the top right, indicating that they were cool but bright. This meant they must be huge, allowing them to shine strongly despite their coolness. They became known as red giants. Some stars appeared on

THE HERTZSPRUNG-RUSSELL DIAGRAM

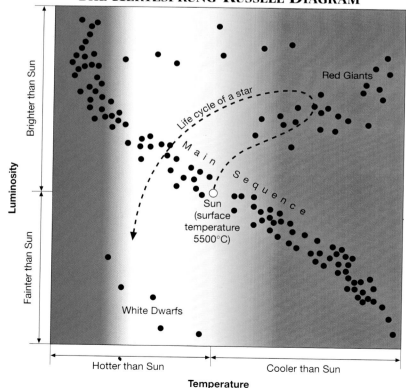

This modern version of the Hertzsprung-Russell Diagram uses colour as well as a scale to indicate the stars' temperatures. The original diagram was in black and white.

on the bottom left, indicating that they were hot but dim. This was because they were very dense but too small to shine brightly. They were called white dwarfs.

Astronomers slowly learned to interpret the graph. They found that a star begins life in the main sequence, where it turns hydrogen into helium to produce energy. When its supply of hydrogen runs out, the star collapses, making its core heat up and outer layers expand. The layers then cool and redden, forming a red giant. Finally, they drift into space, leaving only a core. The star is now a white dwarf and will soon die.

Russell's graph, now known as the Hertzsprung-Russell Diagram (see box), is nothing less than a guide to the evolution of stars. It is still widely consulted by astrophysicists across the world.

Henry Russell came from Oyster Bay, New York. He studied at Princeton University in the USA, then at Cambridge University in England before returning to teach at Princeton. In 1911, Russell became Princeton's Professor of Astronomy. Eleven years later, he was also appointed research associate at Mount Wilson Observatory in California.

EJNAR HERTZSPRUNG

Henry Russell was not the first person to suggest the existence of main sequence, giant and dwarf stars. The Danish astronomer Ejnar Hertzsprung had written articles on the subject in a German journal as early as 1905. For this reason, the graph comparing temperature with luminosity is now known as the Hertzsprung-Russell Diagram. Hertzsprung is also famous for having helped to devise a new and accurate method of calculating the distances of stars from the Earth (see page 10).

Many women carried out important early work in astrophysics. Among them was an Englishwoman, Cecilia Payne-Gaposchkin (right), who studied at Cambridge University. She moved to Harvard University in 1923, becoming Professor of Astronomy in 1956. In this extract from her autobiography (1984), she compares her experiences as a female astronomer in England and the USA.

Payne-Gaposchkin is talking about the 1930s here.

Women [at Cambridge] were segregated in the lecture room. Even in the laboratory they were... treated as second-class students... We manage things better in the United States. Even 50 years ago a woman might do astronomical research and even make a name by publication. She might hold a position – without a title and ill-paid, it is true – and she could meet on equal terms with any astronomer in the world. In my early days at Harvard... How we argued, how we walked about the streets and sat talking in restaurants until the manager turned off the lights in despair! We met as equals; nobody ever condescended to me on account of sex or youth.

EDWIN HUBBLE
AND THE EXPANDING UNIVERSE

By the 20th century, astronomers knew that the Earth was part of the Milky Way galaxy, a vast collection of stars. But they were unsure what lay beyond. In seeking to find out, American astronomer Edwin Powell Hubble uncovered important truths about the nature of the universe.

Hubble originally trained as a lawyer (see box page 11), but in 1914 decided to become an astronomer instead. Eventually, in 1919, he took up a post at the Mount Wilson Observatory in California. Another Mount Wilson astronomer, Harlow Shapley, was already investigating the size and shape of the Milky Way. To do so, he had used a new method of calculating distance devised by Henrietta Leavitt (see box). Some experts believed that Shapley had mistakenly included objects that were not part of the galaxy in the area he had measured, and that as a result his measurements were wrong. In particular, they thought that the spiral nebulae – bright, spiral-shaped clouds of gas and dust – he had included were separate galaxies.

Hubble soon began his own observations of the nebulae to see what they really were. He concentrated on the Andromeda Nebula, peering at it through Mount Wilson's new 100-inch (2.54m) telescope. It revealed that the 'nebula' was in fact made up of stars. Some were Cepheid variables (see box), so Hubble used the Leavitt method to work out Andromeda's distance from the Earth. In 1924, he showed that it was far beyond the possible limits of the Milky Way. In other words it was another, separate galaxy, the first whose existence had been proved.

Hubble soon discovered that more so-called nebulae were actually galaxies. He also noticed that the spectra produced by light from many of them displayed red shift. This means that, compared with a standard light source, the dark lines in their colour bands (see page 8) had shifted towards the red end of the spectrum. This effect occurs when light waves from an object lengthen, and they lengthen when the object is moving away from the Earth. So the greater the red shift, the faster the object is retreating.

By 1929, Hubble had gathered a mass of information about the distances of galaxies from

The existence of the Andromeda Galaxy (centre) was proved by Edwin Hubble in 1924. Its vast distance from Earth – 2.2 million light years – had previously made its identification as a galaxy impossible.

THE SECRETS OF THE CEPHEIDS

The Cepheid variables are so called because the first of these stars was discovered in the constellation Cepheus, and because they pulsate, emitting a variable amount of light. In the early 20th century, American astronomer Henrietta Leavitt studied these stars. She found that the longer a Cepheid took to complete each pulsation, the greater its luminosity. So astronomers can work out a star's luminosity from its pulsation period. By comparing a star's absolute magnitude (its luminosity at a standard distance from Earth) with its apparent magnitude (its luminosity as it appears from Earth), they can also calculate its distance from the Earth. Leavitt's work was developed by Ejnar Hertzsprung (see page 9), Harlow Shapley and others.

the Earth and, using red shifts, about their speed. He was able to show that the more distant a galaxy is, the faster it is moving. In 1929, he published Hubble's Law which expressed this fact in a mathematical formula. His findings proved that the universe is expanding at a rapid rate. They were to help other astronomers calculate the age and possible fate of the universe (see pages 12-15).

RED SHIFT

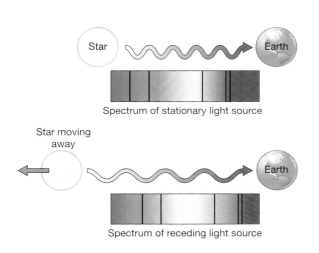

This diagram shows how the light waves from an object lengthen when it is moving away from the Earth, and how the dark bands in the light's spectrum shift towards the red end.

EDWIN POWELL HUBBLE (1889-1953)

Edwin Powell Hubble was born in Marshfield, Missouri, USA. He studied law at university in Chicago, then at Oxford, England. Throughout his student years, he was a keen amateur astronomer as well as a gifted athlete and boxer. On returning to the USA in 1913, Hubble became a lawyer. But a year later he abandoned this career to take a job at the Yerkes Observatory, part of Chicago University. He moved to the Mount Wilson Observatory in 1919 (see main text). There he not only proved that the universe is expanding but devised the first major galaxy classification system. Hubble stayed at Mount Wilson for the remainder of his career.

Edwin Powell Hubble in front of the 2.54m Mount Wilson telescope that he used to study the Andromeda Galaxy.

Astronomers of the 1920s were a close-knit community. This extract is from a letter that Hubble wrote to Harlow Shapley, by then Director of the Harvard College Observatory, when he discovered the first Cepheid variables in the Andromeda Nebula.

In the late 18th century, French astronomer Charles Messier (1730-1817) made three catalogues of all the 'nebulae' that he could see. He gave each one an 'M-number' like this.

February 19, 1924.

...
Dear Shapley:-
You will be interested to hear that I have found a Cepheid variable in the Andromeda Nebula (M31). I have followed the nebula this season as closely as the weather permitted... in the last five months [and found] two variables... last week... Enclosed is a copy of the normal light curve [of the first], which... shows the Cepheid characteristics in an unmistakable fashion. I have a feeling that more variables will be found... Altogether the next season should be a merry one.

GEORGE GAMOW AND THE BIG BANG

Edwin Hubble (see pages 10-11) showed that the universe is expanding. Other scientists, such as Ukrainian-American George Gamow, went on to consider how and when this ever-changing universe began – and where it might end.

Hubble was the first astronomer to find physical evidence of the universe's expansion, in 1924. At about the same time, other astronomers were trying to find theoretical proof of the same phenomenon. They included Georges Edouard Lemaître, a Roman Catholic priest from Belgium. In 1927, he used Einstein's equations (see pages 34-7) to calculate that the universe was expanding. Then he went on to speculate about how that expansion had begun. His theory was that all the matter in the universe had originally been squashed into one incredibly dense 'primeval atom'. This atom, which he also called the 'cosmic egg', had at first slowly disintegrated. But then it had become unstable and violently exploded. As a result, all the matter it contained was still flying apart.

Abbé (Father) Georges Lemaître

 After Georges Lemaître died in 1966, a document headed 'The Expanding Universe' was found in his belongings. It probably dates from the late 1930s or early 1940s. The document clearly explains Lemaître's ideas about the origins of the universe. It also shows that he had considered whether the universe had been created by God. But he clearly states that astronomy cannot answer such questions.

17. THE BEGINNING OF SPACE.
...Physically everything happens as if [the origin of the universe] was really a beginning. The question if it was really a beginning or rather a creation: something starting from nothing, is a philosophical question which cannot be settled by physical or astronomical considerations.

There is a nice way to obtain a universe formed of fresh matter... I mean the primeval atom hypothesis. The universe would have started as an atom... The atom would break into two parts, equal or unequal, each piece would break again until after some two hundred and fifty such processes, matter would be reduced to its actual... state.

Two pages from Lemaître's annotated manuscript

In the 1930s, Lemaître's theory caught the attention of other scientists. Among them was physicist George Gamow, who was by then working at the George Washington University in Washington D.C., USA (see box). As Lemaître had provided no real proof of his theory, Gamow set out to do just that. He worked with a student, Ralph Alpher, to carry out his investigations, and in 1948 the two men published their results. They showed that

Lemaître had been broadly right – the universe had begun with an explosion. But it was caused not by the breakdown of the 'primeval atom', but by the gradual build-up of a hot, dense cloud of gas. Gamow called this giant explosion the Big Bang, the name by which it is still known.

The Big Bang theory seemed likely to be true. However, Gamow still had to work out how the chemical elements that make up the universe could have ➲

GEORGE GAMOW (1904-68)

George Gamow (see picture below) was born in Odessa, Ukraine. His interest in science began when he was still a young child. On his 14th birthday, his father gave him a telescope, and he started to observe the skies regularly. After leaving school, Gamow continued his studies at the University of Leningrad in the USSR before going abroad to carry out research. He met many other brilliant physicists and flourished in the exciting atmosphere of discovery. Gamow studied hard, but found plenty of time for play, too. He was well-known for his late-night partying and practical jokes. Finally, in 1934, Gamow settled in the USA, becoming Professor of Theoretical Physics at the George Washington University, Washington D.C. In 1956 he moved to the University of Colorado, where he remained until his death.

THE BIG BANG

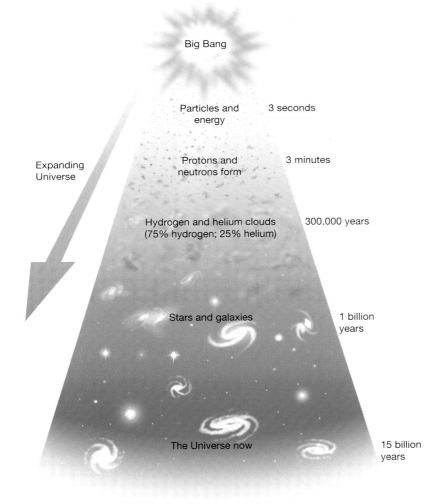

- Big Bang
- Particles and energy — 3 seconds
- Expanding Universe
- Protons and neutrons form — 3 minutes
- Hydrogen and helium clouds (75% hydrogen; 25% helium) — 300,000 years
- Stars and galaxies — 1 billion years
- The Universe now — 15 billion years

The Big Bang caused all the matter in the universe, which had been packed into a super-dense mass, to explode outwards. Gradually the matter took on new forms until the universe we know today was formed.

formed in this process. In the 1950s, he proved how, in the hot, dense early universe, subatomic particles could have combined to produce the two lightest elements, hydrogen and helium. But he also showed that it was impossible for heavier elements to have emerged in this way. Scientists now believe that other elements were created inside stars, which formed from fused hydrogen and helium.

Even stronger evidence for the Big Bang soon appeared. In 1948, Gamow had suggested that the remains of the radiation produced by the explosion must still be in the universe. He had also said that, as the radiation had been cooling ever since, its temperature would now be approximately -270°C. In 1964, American astrophysicists Arno Penzias and Robert Wilson detected this background radiation, which was at about the temperature that Gamow had predicted. For many scientists, this was the final proof that the universe really did begin with the Big Bang.

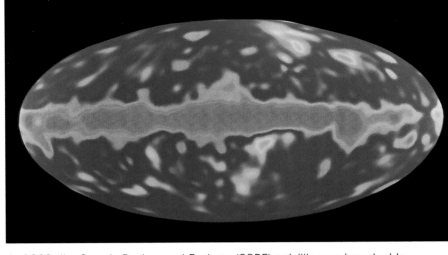

In 1989, the Cosmic Background Explorer (COBE) satellite was launched to measure background radiation. Data from the satellite was used to make this microwave map of the universe. The colours show temperature variations. Dark blue areas are at the radiation's average temperature. Light blue areas are cooler, red and pink areas hotter.

THE STEADY STATE THEORY

In the 1940s, three astronomers strongly opposed Gamow's Big Bang theory. They were Hermann Bondi, Thomas Gold and Fred Hoyle, a British scientist at Cambridge University. In *A New Model for the Expanding Universe* (1948), Hoyle explained his opposing Steady State theory. It says that the Big Bang never happened, but that the universe is expanding because new matter is continually being created. This replaces the matter drifting away into space, and so the universe always looks the same. At first, the Steady State theory had many supporters. But Penzias and Wilson (see main text) provided powerful evidence for the Big Bang and few scientists now accept Hoyle's ideas.

Gamow first gave a detailed explanation of his Big Bang theory in a book, *The Creation of the Universe* (1952). This extract comes from it.

Cosmological means 'relating to the study of the universe'.

The discovery that our universe is expanding provided a master key to the treasure chest of cosmological riddles. If the universe is now expanding, it must have been once upon a time in a state of high compression. The matter which is now scattered through the vast empty space of the universe in tiny portions which are individual stars must at that time have been squeezed into a uniform mass of very high density... At present the possible maximum density of this... matter is not accurately known. The nearest guess is that... each cubic centimeter of space contained at that time a hundred million tons of matter!

Gamow wrote many popular books about science, in which he tried to explain difficult ideas to ordinary readers. The books were enlivened by quirky drawings, some by Gamow himself, as well as by songs and poetry. This poem extract, from *Mr Tompkins in Wonderland*, takes the form of an imaginary argument between two British astronomers. The first, Martin Ryle, helped to prove the Big Bang theory using information collected by radio telescope. The second, Fred Hoyle, was the main originator of the Steady State theory.

FROM BEGINNING TO END

Using Hubble's Law and his own observations, Gamow calculated that the Big Bang took place about 17,000 million years ago, so beginning the universe. Modern scientists think that this is approximately correct. It is impossible to know exactly how the universe will develop in the future, but there are three main options. The universe may be 'flat', which means it will use up all its energy and come to a standstill. It may be 'closed', which means it will stop expanding and contract until all its matter comes together again in the 'Big Crunch' (see page 16). However, the most recent findings suggest that the universe is probably 'open', which means it will continue to expand forever.

Part of the poem shown on the right as it appears in Gamow's book.

COSMIC OPERA

one million light-years away—show how it actually looked one million years ago. Thus, what Ryle sees, or should I rather say hears, through his radio-telescope, corresponds to the situation which existed in that distant part of the universe many thousand millions of years ago. If the universe were really in a steady state, the picture should be unchanged in time, and very distant galaxies as observed from here now should be seen distributed in space neither more densely nor rarely than the galaxies at shorter distances. Thus Ryle's observations showing that distant galaxies seem to be more closely packed together in space is equivalent to the statement that the galaxies everywhere were packed more closely together in the distant past of thousands of millions of years ago. This contradicts the steady state theory, and supports the original view that the galaxies are dispersing and that their population density is going down. But of course we must be careful and wait for further confirmation of Ryle's results.'

'By the way,' continued the professor, extracting a folded piece of paper from his pocket, 'here is a verse which one of my poetically inclined colleagues wrote recently on this subject.' And he read: 'Your years of toil,'

Said Ryle to Hoyle,
'Are wasted years, believe me.
The steady state
Is out of date.
Unless my eyes deceive me,

My telescope
Has dashed your hope;
Your tenets are refuted.
Let me be terse:
Our universe
Grows daily more diluted!'

Said Hoyle, 'You quote
Lemaître, I note,
And Gamow. Well, forget them!
That errant gang
And their Big Bang—
Why aid them and abet them?

63

THE WOODCARVER

'This is our large cyclotron or "atom-smasher"'

Your tenets are refuted means 'your beliefs have been disproved'.

See box on page 14.

This lively drawing from another of Gamow's books shows a particle accelerator. You can see a modern example on page 33.

'Your years of toil,'
Said Ryle to Hoyle,
'Are wasted years, believe me.
The steady state
Is out of date.
Unless my eyes deceive me,

My telescope
Has dashed your hope;
Your tenets are refuted.
Let me be terse:
Our universe
Grows daily more diluted!'

Said Hoyle, 'You quote
Lemaître, I note,
And Gamow. Well, forget them!
That errant gang
And their Big Bang—
Why aid them and abet them?

You see, my friend,
It has no end
And there was no beginning,
As Bondi, Gold,
And I will hold
Until our hair is thinning!'

STEPHEN HAWKING
AND BLACK HOLES

Towards the end of their lives, many stars turn into red giants, then white dwarfs (see pages 8-9). However, stars with a mass at least three times greater than that of the Sun (see box page 17) become black holes. These are among the most mysterious objects in space, but thanks to the work of Stephen Hawking and other scientists, they are slowly revealing their secrets.

In the 18th century, the astronomers John Michell and Pierre-Simon de Laplace were already wondering if there might be 'dark stars' from which light could not escape. As such stars would be invisible, these men could not prove their theory. But 20th-century scientists had other ways of finding out the truth. After Albert Einstein published his general theory of relativity in 1916 (see page 36), scientists used his equations to suggest that, once stars above a certain mass had collapsed, their gravitational fields would prevent light from emerging.

In 1965, British physicist Roger Penrose proved this theory and showed that these stars collapsed to form a singularity, a point where matter is so dense that it takes up no space. As even light cannot escape from a singularity, a circular dark area is created around its edge. In 1967, Professor John Wheeler of Princeton University in the USA coined a name for this 'collapsed star' phenomenon – black hole.

Penrose's research attracted the interest of another physicist, the Oxford-educated Englishman Stephen Hawking (see box). He was fascinated by cosmology (the study of the universe), and particularly by the origins of the universe. Research inspired by Penrose's findings led Hawking to write a revolutionary PhD thesis in 1965. It stated that, as a star can collapse inwards to form a black hole, so a black hole can explode outwards. Such a process, Hawking showed, had caused the Big Bang. If it reversed again, it would lead to the Big Crunch.

Since 1965, Hawking has continued to investigate black holes. In particular, he has shown that they are not really 'black' at all, as radiation can emerge from them. He has also suggested that dense 'mini-black holes' formed just after the Big Bang, but none has yet been detected. Hawking continues to devise complex new theories that are helping to unravel the mysteries of the early universe.

STEPHEN HAWKING (1942-)

Stephen William Hawking (see picture below) was born in Oxford, and studied at the city's university before going to Cambridge University. In the early 1960s, Hawking started to have difficulty co-ordinating his movements. Doctors found he had motor neurone disease, which causes muscle wasting and paralysis. Despite this shocking diagnosis, Hawking completed his PhD thesis at Cambridge in 1965 (see main text), then remained there as an academic. The progress of his disease eventually left him unable to walk or talk, so he began to use a wheelchair and a voice synthesizer to 'speak' his words. The career of this highly gifted man nevertheless flourished. In 1979, he became Lucasian Professor of Mathematics at Cambridge, a post he still holds today.

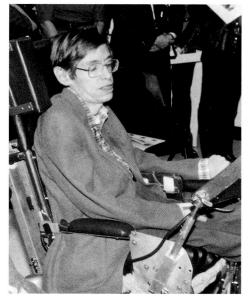

Roger Penrose is Rouse Ball Professor of Mathematics at Oxford University, where he is developing complex new ideas about the universe, known as twistor theory. He has also written a book with Stephen Hawking called *The Nature of Space and Time* (1996).

In 1965, astronomers detected a strong but invisible X-ray source in the constellation Cygnus. Christened Cygnus X-1, it was located near a supergiant star. Hawking believes that the X-rays result from the presence of a black hole whose gravity is pulling gas away from the star. This gas forms an accretion disc around the hole (see illustration above right). This accretion disc then emits X-rays that astronomers can both measure and photograph (see picture right).

A BLACK HOLE

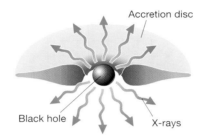

Accretion disc

Black hole

X-rays

A BLACK HOLE IS BORN

When a massive star begins to run out of hydrogen (see page 9), it forms not a red giant, but a red supergiant. Nuclear reactions then take place inside this huge star, forming spheres of different elements around its core. Eventually, most of the core itself is turned into iron. This causes the star to collapse, then explode as a supernova. In the process, the star's outer layers are hurled into space. If the mass of the remaining core is between 1.4 and 3 times that of the Sun, it shrinks to form a dense neutron star. If its mass is any greater, it forms a black hole.

In 1988, Stephen Hawking published *A Brief History of Time*, in which he used simple language to explain cosmology past and present. This extract describes what might happen to an astronaut inside a black hole.

If an astronaut falls into a black hole, its mass will increase, but eventually the energy equivalent of that extra mass will be returned to the universe in the form of radiation... Thus, in a sense, the astronaut will be "recycled". It would be a poor sort of immortality, however, because any personal concept of time for the astronaut would almost certainly come to an end as he was torn apart inside the black hole!

CHAPTER 2

CHEMISTS

JOHN DALTON
AND ATOMIC THEORY

Everything, whether a solid, a liquid or a gas, is made of matter. But until the 17th century, scientists had little idea of what matter itself was made of. One of the major breakthroughs in understanding was made by John Dalton (see box).

Ancient Greeks thought matter was made up of four elements – earth, air, fire and water. Some believed elements were made of atoms (indivisible particles). During the Middle Ages, superstitious alchemists did little to further scientific knowledge. But, gradually, scientists began to test their ideas about matter and to come up with new theories. *The Sceptical Chymist* (1661) by Irishman Robert Boyle contained the first modern definition of an element as a substance that cannot be broken down into a simpler one. It also revived the idea that elements were made of 'primary particles'.

Chemists gradually began to work out which substances were elements according to Boyle's definition and which were compounds – chemically combined groups of elements. For example, in the 18th century the English scientist Henry Cavendish showed that water was a compound of the elements hydrogen and oxygen, so could not itself be an element. However, less progress was made in investigating 'primary particles'.

In the late 18th century, Englishman John Dalton began his enquiries. Dalton's interest in the make-up of matter was prompted by his study of meteorology. As weather is affected by the air, he decided to investigate nitrogen and oxygen, the main elements in air. His tests showed these gases were not combined in a compound but in a mixture, where they acted independently. For example, they exerted different amounts of pressure and one was more easily absorbed by water vapour than the other. Tests on other gases showed that they too had different levels of solubility in water, a sign that they also had different weights.

These experiments led Dalton to believe that every element was made up of a different type of particle, each with a different weight. Like the Greeks, Dalton called these particles atoms. In 1803, he announced his ideas to the Manchester Literary and Philosophical Society, and in 1808 published a fuller account in *A New System of Chemical Philosophy* (see document). Dalton's theory, later modified by Italian Amedeo Avogadro (see page 20), was the foundation of modern science.

JOHN DALTON (1766-1844)

John Dalton (see picture below) was born into a Quaker family in Eaglesfield, Cumbria. When he was 12, he became a teacher at the local school. At 15, he took up a post at another school in nearby Kendal, where he also started to teach himself science. In 1793, Dalton became a professor at New College, Manchester, and a year later was elected to the Manchester Literary and Philosophical Society. Its members met regularly to give and hear lectures about the latest ideas. Dalton later presented many of his own scientific ideas there, as well as at the Royal Institution in London (see page 28). He was made a fellow of London's Royal Society, an older and more prestigious organisation than even the Institution, in 1822.

COLOUR BLINDNESS

John Dalton was colour blind and carried out many experiments to discover more about this condition. For example, he recorded how silk threads of different colours appeared to him in daylight and in candle light (see right). In 1794, he presented a paper on the subject to the Manchester Literary and Philosophical Society. *Extraordinary Facts Relating to the Vision of Colours* contained the first ever scientific description of colour blindness, which is now sometimes known as daltonism.

Dalton attached the silks he used to test his colour blindness to the left-hand pages of this notebook, and wrote down how the colours appeared to him on the right-hand pages.

The title page of John Dalton's book

A

NEW SYSTEM

OF

CHEMICAL PHILOSOPHY.

PART II.

BY

JOHN DALTON.

Manchester:
Printed by Russell & Allen, Deansgate,
FOR
R. BICKERSTAFF, STRAND, LONDON,
1810.

In *A New System of Chemical Philosophy* (1808) Dalton explained all the essential ideas of his atomic theory. He stated that all matter is made of indivisible particles called atoms, and that the atoms of each element are of a different type and atomic weight. He also wrote that atoms are rearranged in chemical reactions, and may combine to form compounds. Finally, as this extract shows, he declared that atoms cannot be created or destroyed.

This means 'the separation and combination of atoms that take place in chemical reactions'.

Chemical analysis and synthesis go no farther than to the separation of particles one from another, and to their reunion. No new creation or destruction of matter is within the reach of chemical agency. We might as well attempt to introduce a new planet into the solar system, or to annihilate one already in existence, as to create or destroy a particle of hydrogen. All the changes we can produce, consist in separating particles that are in a state of cohesion or combination, and joining those that were previously at a distance.

DMITRI MENDELEEV
AND THE PERIODIC TABLE

As the 19th century progressed, scientists slowly learned more about atoms. They also discovered more elements until, by the 1860s, they knew of 60. But it was Russian chemist Dmitri Ivanovich Mendeleev who found out how all these elements were related.

In the early 19th century, Italian scientist Amedeo Avogadro showed that the smallest naturally occurring particles of an element are not usually individual atoms but molecules. He also showed that the same volume of any gas, at the same temperature and pressure, contains the same number of molecules. This finding, known as Avogadro's Law, provided scientists with a reliable way of calculating the relative weights of molecules and, eventually, of atoms too.

Some scientists then began to think that the new atomic weights could be used to classify the elements. In 1864, English chemist John Newlands made one of the first attempts to do this. He noticed that when the known elements were listed in order of their atomic weights, every eighth one had similar properties. He called this the Law of Octaves. Newland's discovery led him to prepare a table in which every eighth element began a new row. The finished layout had seven columns, and the elements in each column shared many important characteristics.

Newlands, however, made one significant error. He failed to consider that some elements might not yet have been discovered – so he left no room for them in his table. It took the insight of Dmitri Mendeleev (see box) to take this next step forward. In February 1869, to help order his thoughts for writing a chemistry textbook, he wrote the name of each element and its properties on a card. Then he moved the cards around until he found the layout that best showed the relationships between them. Like Newlands, he put elements with similar properties in the same columns. But if no known element was suitable for a particular slot, he left a gap. Mendeleev also predicted the properties of the elements that he believed would one day fill the gaps.

By the time he had finished, Mendeleev had created the Periodic Table, so called because there are regular periods (intervals) between elements with similar properties. Three of the 'missing' elements were soon found, proving the table's worth. It has since been much adapted and expanded (see box page 21), but remains a vital scientific tool.

DMITRI IVANOVICH MENDELEEV (1834-1907)

Dmitri Ivanovich Mendeleev (see picture left) was born in Tobolsk, in Siberia, Russia. While he was still a teenager his father died, so his mother had to bring up her family of 14 children alone. She wanted Dmitri, her youngest and her favourite, to get a good education, so sent him to the Central Pedagogic Institute in the city of St Petersburg. Mendeleev repaid his mother's faith in him when, in 1866, he became Professor of Chemistry at St Petersburg University. It was there that he devised the Periodic Table, as well as writing a major textbook, *The Principles of Chemistry* (1868-70). Mendeleev continued to teach at St Petersburg for many years. But his support for political change in Russia led him to resign in 1890.

The document below is Mendeleev's first, handwritten version of the Periodic Table. The earliest published Periodic Table appeared in *On the Relation of the Properties to the Atomic Weights of Elements* (1869). At the bottom of the page, you can see how the Periodic Table looks today.

CHANGING TABLES

As knowledge grew, the Periodic Table changed. In the 20th century, scientists discovered that atoms are made up of subatomic particles (see pages 32-33), among them protons. They also found that the atoms of each element contain a different number of protons. Now the elements in the table are arranged according to these atomic numbers, from lowest to highest. The new layout is similar to Mendeleev's because atomic weights, which he used to order his table, rise with atomic number. The number of elements in the table has changed, too. Mendeleev included only about 60. Now there are at least 118.

American chemist Glenn Seaborg helped to discover eight new elements between 1940 and 1958. They included Element 101, which was given the name mendelevium, after Mendeleev.

THE PERIODIC TABLE

Each square in this modern Periodic Table contains the atomic number of the element, its chemical symbol and full name. Elements that are related, for example particular types of metal or gas, all have squares of the same colour.

- Alkali metals
- Alkaline-earth metals
- Transition metals
- Lanthanoids
- Actinoids
- Poor metals
- Semi-metals
- Non-metals
- Noble gases

hydrogen 1 H																	helium 2 He
lithium 3 Li	beryllium 4 Be											boron 5 B	carbon 6 C	nitrogen 7 N	oxygen 8 O	fluorine 9 F	neon 10 Ne
sodium 11 Na	magnesium 12 Mg											aluminium 13 Al	silicon 14 Si	phosphorus 15 P	sulphur 16 S	chlorine 17 Cl	argon 18 Ar
potassium 19 K	calcium 20 Ca	scandium 21 Sc	titanium 22 Ti	vanadium 23 V	chromium 24 Cr	manganese 25 Mn	iron 26 Fe	cobalt 27 Co	nickel 28 Ni	copper 29 Cu	zinc 30 Zn	gallium 31 Ga	germanium 32 Ge	arsenic 33 As	selenium 34 Se	bromine 35 Br	krypton 36 Kr
rubidium 37 Rb	strontium 38 Sr	yttrium 39 Y	zirconium 40 Zr	niobium 41 Nb	molybdenum 42 Mo	technetium 43 Tc	ruthenium 44 Ru	rhodium 45 Rh	palladium 46 Pd	silver 47 Ag	cadmium 48 Cd	indium 49 In	tin 50 Sn	antimony 51 Sb	tellurium 52 Te	iodine 53 I	xenon 54 Xe
caesium 55 Cs	barium 56 Ba	57-71	hafnium 72 Hf	tantalum 73 Ta	tungsten 74 W	rhenium 75 Re	osmium 76 Os	iridium 77 Ir	platinum 78 Pt	gold 79 Au	mercury 80 Hg	thallium 81 Ti	lead 82 Pb	bismuth 83 Bi	polonium 84 Po	astatine 85 At	radon 86 Rn
francium 87 Fr	radium 88 Ra	89-103	rutherfordium 104 Rf	dubnium 105 Db	seaborgium 106 Sg	bohrium 107 Bh	hassium 108 Hs	meitnerium 109 Mt	ununnilium 110 Uun	unununium 111 Uuu	ununbium 112 Uub		ununquadium 114 Uuq		ununhexium 116 Uuh		ununoctium 118 Uuo

lanthanum 57 La	cerium 58 Ce	praseodymium 59 Pr	neodymium 60 Nd	promethium 61 Pm	samarium 62 Sm	europium 63 Eu	gadolinium 64 Gd	terbium 65 Tb	dysprosium 66 Dy	holmium 67 Ho	erbium 68 Er	thulium 69 Tm	ytterbium 70 Yb	lutetium 71 Lu
actinium 89 Ac	thorium 90 Th	protactinium 91 Pa	uranium 92 U	neptunium 93 Np	plutonium 94 Pu	americium 95 Am	curium 96 Cm	berkelium 97 Bk	californium 98 Cf	einsteinium 99 Es	fermium 100 Fm	mendelevium 101 Md	nobelium 102 No	lawrencium 103 Lr

MARIE CURIE
AND RADIOACTIVITY

Several of the elements discovered shortly after Mendeleev's creation of the Periodic Table (see page 20) shared a special property. This was radioactivity, the emission of particles or rays caused by the spontaneous disintegration of atomic nuclei (see page 32). This phenomenon was investigated by many scientists. Among the most dedicated was a young Polish woman, Marie Sklodowska Curie.

In 1895, German physicist Wilhelm Röntgen discovered X-rays (see page 30). His work was noticed by French scientist Antoine Henri Becquerel,

Professor of Physics at the Ecole Polytechnique in Paris. Becquerel was already studying fluorescence, the radiation given off by some substances when they are exposed to the Sun's rays. He now began to wonder if the fluorescence from one of these substances, potassium uranyl sulphate, might also contain X-rays.

In 1896, Becquerel tested his theory. He knew that X-rays discoloured photographic plates. So he placed some potassium uranyl sulphate crystals in the sunshine to make them fluoresce, then put a photographic plate wrapped in black paper next to

them. He knew that the Sun's rays could not get through the black paper, but that any X-rays given off by the crystals could. When the photographic plate darkened, he believed that this was a result of the X-rays, and that this proved they were present in the fluorescence.

Then something unexpected happened. The weather grew cloudy and Becquerel had to delay his experiments. He put the wrapped photographic plate and crystals away in a drawer. When he went to retrieve them, he found that the plate was blurred, even though the crystals had not been fluorescing. The conclusion

MARIE CURIE (1867-1934)

Marie Curie was born in Warsaw, Poland. In 1891, she left Poland to study science at the Sorbonne, part of Paris University. Four years later, she married another scientist, Frenchman Pierre Curie. In 1904, a year after the couple had won the Nobel Prize (see main text), Pierre became Professor of Physics at the Sorbonne. Marie meanwhile looked after their two daughters and worked part-time at a girls' school. Tragically, in 1906, Pierre was knocked over and killed by a cart. Marie was then appointed to his former post, becoming the first woman professor at the Sorbonne. In 1914 she became head of the Paris Institute of Radium. Unaware of the dangers of radioactivity, Marie never protected herself from the radioactive elements with which she worked. The prolonged exposure eventually caused her to develop leukaemia, from which she died.

This illustration of Pierre and Marie Curie appeared on the cover of a French magazine in 1904.

eventually arrived at by Becquerel was that another form of radiation was being produced. He suggested the source was uranium, an element in the crystals.

Now Marie Curie, who was also living in Paris, began to take an interest. By the time Becquerel's findings became public, she was about to begin research for a thesis. The intriguing new form of radiation that Becquerel had discovered seemed an ideal subject. Curie began by measuring the rays given off by uranium. She then tested other elements and found that one, thorium, emitted the same type of radiation. Curie gave this radiation a name – radioactivity. Next, she tested uranium and thorium compounds. The amount of radioactivity emitted was not affected by the other elements in them, or by varying conditions. It depended only on the amount of uranium or thorium present. So she concluded that radioactivity was a property of their atoms. Later investigations proved her right. ➲

Marie Curie's elder daughter, Irène, became a scientist like her mother. But it was Marie's younger daughter, Eve, who wrote her biography, *Marie Curie* (1938). The two extacts below show how Marie juggled her work as a scientist with her life as a mother. The first, from her personal notebook, is dated August 1898. The second, from the proceedings of the French Academy of Science, appeared in December the same year.

Irène has cut her seventh tooth, on the lower left. She can stand for half a minute alone. For the past three days we have bathed her in the river. She cries, but to-day (fourth day) she stopped crying and played with her hands in the water. She plays with the cat and chases him with war cries. She is not afraid of strangers any more. She sings a great deal...

The various reasons we have just enumerated lead us to believe that the new radioactive substance contains a new element to which we propose to give the name of RADIUM. The new radioactive substance certainly contains a very strong proportion of barium; in spite of that its radioactivity is considerable. The radioactivity of radium, therefore, must be enormous.

WAR WORK

From 1914 to 1918, the First World War raged across large areas of north-western France. In this time of need, Marie Curie put her scientific knowledge to practical use. She realised that X-rays would be a great help to surgeons who were treating wounded soldiers. So she collected X-ray machines from university laboratories and sent them to hospitals in and around Paris. It quickly became clear that doctors would need X-ray equipment in the war zones, too. So Curie installed the necessary apparatus into about 20 vans, which became known as 'little Curies'. One of them often ferried Curie herself out to help battlefield casualties.

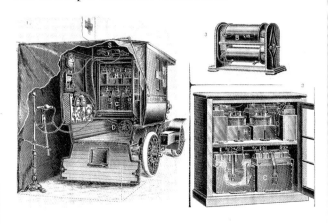

A cut-away image of a French X-ray van, showing the special equipment that it carried.

Barium is a non-radioactive metallic element.

As a next step, Curie began to investigate pitchblende, a uranium-rich mineral. When she measured its radioactivity, she found that it was higher than that of pure uranium, so pitchblende had to contain at least one other intensely radioactive element. Her husband, Pierre (see box page 22), abandoned his own work to help his wife search for this substance. In 1898, the couple found two new radioactive elements. The first was named polonium, after Marie's homeland. The second, even more radioactive, was christened radium.

The Curies now set about processing several tonnes of pitchblende, with the aim of extracting pure radium. They set up a laboratory in a schoolyard shack and worked doggedly for four years. By 1902, they had just one-tenth of a gram, but this was enough for another chemist to work out its atomic weight. In 1903, Marie finished her thesis. That same year, she, her husband and Becquerel won the Nobel Prize for Physics, for their combined studies of radioactivity. In 1911, Marie alone won the Nobel Prize for Chemistry, for her discovery of radium and polonium.

Marie Curie's discovery of radioactivity has made a lasting impact, for good and ill. Among the

This haunting image shows Marie Curie at work. She achieved her results with basic apparatus rather than the vast array of complex equipment available to modern scientists.

EXTRA ELEMENTS

In the early 20th century, scientists following in Marie Curie's footsteps discovered the existence of several more radioactive elements. In 1900, Frenchman André Debierne, once the Curies' assistant at the Sorbonne, found actinium in the mineral pitchblende. German physicist Friedrich Dorn isolated radon, a radioactive gas, in the same year. In 1917, an English and a German research team independently discovered another element, protactinium. It was also found in pitchblende.

In the past, as now, the price of fame was often a loss of privacy. In this letter, also taken from *Marie Curie* (see page 23), Pierre Curie tells a friend how his life has changed since the discovery of radium.

...You have seen this sudden fad for radium. This has brought us all the advantages of a moment of popularity; we have been pursued by the journalists and photographers of every country on earth; they have even gone so far as to reproduce my daughter's conversation with her nurse and to describe the black-and-white cat we have at home. Then we have received letters and visits from all the eccentrics, from all the unappreciated inventors... We have had a large number of requests for money. Last of all, collectors of autographs, snobs, society people and sometimes even scientists come to see us... With all this, there is not a moment of tranquillity in the laboratory...

major benefits are medical applications. For example, radioactive substances can be injected into patients and tracked through their bodies so that doctors can find out what is wrong, and radioactive rays can be directed at cancers to destroy them. However, experts have also learned that exposure to radioactivity can kill – Curie herself was a victim (see box page 22). As a result, many people now consider other applications, such as the use of radioactive fuels in nuclear power stations, and the resulting risk of radioactive leaks, too dangerous. Everywhere scientists are learning to employ radioactivity with care.

The Paris Institute of Radium in 1955. This picture was taken more than 20 years after the death of its remarkable founder.

The Institute of Radium in Paris opened in 1914. It had two laboratories, one for academic study, the other for research into Curietherapy – medical treatments using radioactivity.
Unfortunately, it had very little radium. In 1920, Marie Curie explained the problem to a visiting American, Mrs William Meloney. Back at home, Mrs Meloney raised funds to buy

a precious gram of radium for the institute. She was so successful that in 1921 Curie was able to collect the radium from President Warren G. Harding in Washington D.C. This extract from Eve Curie's biography is Mrs Meloney's own account of her first meeting with Marie Curie.

I tried to explain that American women were interested in her great work, and found myself apologising for intruding upon her precious time. To put me at my ease Mme Curie began to talk about America.

"America," she said, "has about fifty grammes of radium. Four of them are in Baltimore, six in Denver, seven in New York." She went on, naming the location of every grain.

"And in France?" I asked.

"My laboratory has hardly more than a gramme." ...

"If you had the whole world to choose from," I asked impulsively, "what would you take?"

It was a silly question perhaps, but as it happened, a fateful one.

DOROTHY HODGKIN
AND X-RAY CRYSTALLOGRAPHY

In the early 20th century, scientists were struggling to find out more about atoms and molecules. But they were hindered by the fact that they could not see these tiny particles. Then, in 1912, a method of examining them using X-rays was devised. This technique was put to brilliant use by British chemist Dorothy Crowfoot Hodgkin.

X-rays were discovered in 1895 (see page 30). In 1912, as part of his work to reveal their precise nature, German physicist Max von Laue fired X-ray beams into crystals. (Crystals are solids with their atoms arranged in lattices – see diagram.) Von Laue knew that the X-rays would hit the atoms and be diffracted (scattered). But he was thrilled to find that, because of the special arrangement of crystal atoms, the scattered rays made regular patterns on photographic plates. Two British scientists, William Henry and William Lawrence Bragg, realised that the patterns produced by the X-rays were a guide to the three-dimensional structures of the crystals. W.L. Bragg worked out a means of interpreting them, and X-ray crystallography was born.

The technique developed steadily over the following years, and the young Dorothy Hodgkin (see box) made it her speciality. Then, in the mid-1930s, she employed it to investigate insulin. This protein, discovered in 1921, is used in the treatment of diabetes, a disorder caused by the body's inability to process sugars properly. Doctors wanted to understand more about the structure of insulin, and Hodgkin produced crystals suitable for analysis in 1935. But it was many years before she understood their internal form.

In the Second World War, Hodgkin investigated penicillin (see pages 54-5), which by the 1940s

A SODIUM CHLORIDE CRYSTAL

Chlorine atoms   Sodium atoms

This diagram of a sodium chloride (common salt) crystal shows clearly how the atoms of the two elements are arranged in a lattice.

DOROTHY CROWFOOT HODGKIN (1910-94)

Dorothy Mary Crowfoot (see picture below) was born in Cairo, Egypt. When the First World War broke out in 1914, she returned to England with her mother. During the early 1920s, she read two children's lectures about X-ray crystallography by William Henry Bragg (see document) and her lifelong interest began. In 1928, Crowfoot went to Somerville College, Oxford to study chemistry. After graduating, she carried out crystallography research at Cambridge University, before returning to Oxford as a lecturer. In 1937 she married Africa expert Thomas Hodgkin (hence the name by which she is known). Dorothy Hodgkin left Oxford in 1970 to become Chancellor of Bristol University.

was in use as an antibiotic. She revealed its structure in 1945. Three years later, she began to examine Vitamin B12. It was also of medical interest, as people who cannot absorb the vitamin develop pernicious anaemia (a serious red blood cell deficiency). Computers helped to build up a picture of the highly complex molecules that make up B12. It was ready in 1956, and Hodgkin won the Chemistry Nobel Prize for her work in 1964. Hodgkin and her research team made one more breakthrough. In 1969, after 34 years, the structure of insulin was finally discovered.

INTERNATIONAL UNDERSTANDING

Four of Dorothy Hodgkin's uncles were killed in the First World War, and she also witnessed the sufferings of many people in the Second World War. She became dedicated to world peace and was a member of the Campaign for Nuclear Disarmament. She promoted international understanding by visiting scientists in many countries. In China, she was delighted to be shown an insulin structure map made in 1971, just two years after she had completed her own.

William Henry Bragg (see document below) and his son shared the Nobel Prize for Physics in 1915.

Lecture series for children were regular events at the Royal Institution in London from 1826 (see page 29). In 1923 and 1925, the lectures were given by William Henry Bragg. They were later read by the young Dorothy Crowfoot and set her on the path to her future career. This description of X-ray crystallography comes from the 1925 lecture, *Concerning the Nature of Things*.

The discovery of the X-ray has provided means by which we can look far down into the structure of solid bodies, and observe in detail the design of their composition... How far our new powers will carry us, we do not yet know; but it is certain that they will take us far and give us a new insight... Broadly speaking, the discovery of X-rays has increased the keenness of our vision a thousand times, and we can now 'see' the individual atoms and molecules.

The work of an X-ray crystallographer is difficult and time-consuming. But the achievement of a final result can be exhilarating, as this quotation shows. It is part of an article by Dorothy Hodgkin that appeared in the *British Medical Journal* in 1971.

X-rays are diffracted by subatomic particles called electrons (see page 32). By looking at the patterns that the diffractions produce, crystallographers can see how densely the electrons are packed together and so identify the positions of atoms.

I used to say that the evening I developed the first X-ray photograph I took of insulin in 1935 was the most exciting moment of my life. But the Saturday afternoon in late July 1969, when we realized the insulin electron density map was interpretable, runs that moment very close.

CHAPTER 3

PHYSICISTS
MICHAEL FARADAY
AND ELECTROMAGNETIC INDUCTION

By 1800, scientists knew several basic facts about electricity. They were aware that it could be static or flow in a current, and have a positive or negative charge. But understanding of what electricity was, or how its power might be harnessed, was limited. Then, in the 1820s, Englishman Michael Faraday began to find out more.

Other scientists had already begun to investigate electricity in earnest. In 1820, Danish physicist Hans Christian Oersted had shown that when a magnetic compass is placed near a wire with electricity flowing through it, the compass needle moves. This indicated that electric current produced a magnetic field. The French scientist André Marie Ampère went on to discover much more about the link between electricity and magnetism.

Faraday made his first major electrical discovery in 1821, when he suspended a current-carrying wire above a fixed magnet. The magnetic field caused the wire to rotate – he had discovered the motor effect. This led to the development of the electric motor, which today powers everything from hairdryers to cars. However, Faraday was more interested in his theory that, just as an electric current can produce a magnetic field, a magnetic field can produce an electric current. In 1831, he set up an experiment to test his idea. As a first step, he wound two wire coils around an iron ring, then joined one coil to a battery, the other to a

MICHAEL FARADAY (1791-1867)

Michael Faraday was born in Newington, near London. He left school at 13, and was employed as an apprentice bookbinder. Eager for knowledge, Faraday read parts of the books he bound, including an article on electricity in the *Encyclopaedia Britannica*. He was fascinated by what he learned. In 1812, he went to four lectures at the Royal Institution, a scientific organisation in London. The lecturer was famous chemist Humphry Davy. Davy's lectures made Faraday determined to follow a scientific career. So, in 1813, he got a job as one of Davy's assistants at the Royal Institution. In 1825, he became the institution's laboratory director and in 1833 its Professor of Chemistry. Faraday's work in the field of chemistry was important. He discovered benzene in 1825 and published his laws of electrolysis during the 1830s. However, his research into electricity was even more significant (see main text). Faraday worked in his laboratory at the Royal Institution until his death.

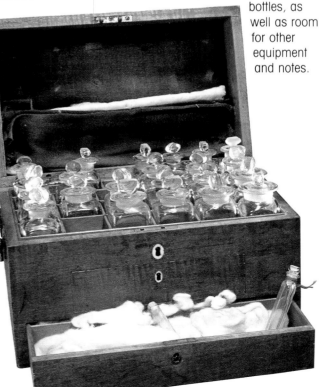

The wooden chemical chest used by Michael Faraday. It has slots for 24 bottles, as well as room for other equipment and notes.

galvanometer (an instrument for measuring electric current).

The result was clear. When the battery was connected, electricity flowed in both coils. There could be only one explanation. The battery had made current flow in the coil to which it was joined. The magnetic field created by this current had then produced a current in the other coil. Faraday's theory had been correct. He later showed a current could be created simply by moving a magnet in and out of a wire coil. This effect, electromagnetic induction, is how power-station generators now produce electricity. Without it, our modern world could not exist.

FORCE FIELDS

Faraday was above all an imaginative practical scientist. But he did propose an important theory about how magnetism and electricity work and affect each other. In 1831, he suggested that magnets were surrounded by invisible lines of force that allowed them to have an effect on objects at a distance. In 1837, he proposed that current-carrying wires were also surrounded by such lines or fields of force. James Clerk Maxwell (see pages 30-1) used advanced mathematics to develop Faraday's ideas.

Michael Faraday gives one of his Christmas lectures to an audience of curious children – and a few adults – at the Royal Institution (see document).

From the 1820s, Faraday organised talks at the Royal Institution especially for non-scientists. He also founded the Royal Institution Christmas lectures for children.
The best-known Christmas lecture series that Faraday gave was _The Chemical History of a Candle_ (1860). This extract is taken from the introduction to the first of these six talks.

Natural philosophy is an old term for 'physical sciences', that is sciences that do not deal with living things.

Faraday is saying that careful examination of even a simple burning candle can open up the whole world of science to a child.

By **seniors**, Faraday means adults. He is telling the children ('juveniles') in his audience that he is speaking to them like a child himself, because he shares their childlike enthusiasm for the subject.

There is no better, there is no more open door by which you can enter into the study of natural philosophy, than by considering the physical phenomena of a candle... before proceeding, let me say this also – that though our subject be so great, and our intention that of treating it honestly, seriously, and philosophically, yet I mean to pass away from all those who are seniors amongst us. I claim the privilege of speaking to juveniles as a juvenile myself...

JAMES CLERK MAXWELL
AND ELECTROMAGNETIC THEORY

James Clerk Maxwell (1831-79) was an exceptionally brilliant Scottish physicist. He made great advances in many fields, particularly the study of gases, colour vision and optics. But his most revolutionary work transformed scientific understanding of electricity, magnetism and light.

As a young science student and lecturer, Maxwell studied the work of Michael Faraday, in particular his lines of force theory (see page 29), for some years. In 1856, he finished a paper that explained the results of Faraday's electromagnetic experiments in mathematical terms. By 1860, when he became Professor of Natural Philosophy and Astronomy at King's College, London, Maxwell was ready to continue his research in this field.

Maxwell assembled his theories over the next four years. Then, in 1864, he published his greatest work, called *A Dynamical Theory of the Electromagnetic Field*. It contained four equations, known as Maxwell's equations, that explained almost everything about electricity, magnetism and the relationship between them. In particular, the equations showed that electric and magnetic fields move along together as electromagnetic waves, a type of radiation. The fields occur at right angles to each other, but cannot be separated (see diagram).

The book contained other ground-breaking ideas, too. Maxwell had discovered that electromagnetic waves travel at the speed of light (about 300,000 km per second). So he concluded that light itself was a type of electromagnetic radiation. Maxwell's calculations also showed that there were probably other forms of electromagnetic radiation with different wavelengths. He was proved correct after his death, when radio waves (see box page 31), X-rays (see picture), gamma, ultraviolet and microwaves were all discovered. Together with the previously discovered infrared waves they make up the electromagnetic spectrum.

James Clerk Maxwell's theoretical genius has had numerous practical effects that he could never have foreseen. Today, gamma rays are used for some types of cancer therapy, while X-rays allow medical staff to see inside the body (see picture below). Ultraviolet rays from sunbeds produce artificial tans, although skin experts now consider this process to be dangerous. Infrared photography is used for medical diagnosis, the location of underground minerals and much more. Microwave ovens cook food quickly and efficiently, while radio waves carry sound and television pictures into our homes.

In 1895, German physicist Wilhelm Röntgen was investigating cathode rays (see page 32) by projecting them on to a screen. Then he noticed that another, unknown form of rays was causing a more distant screen to fluoresce. He had discovered X-rays, which in 1912 were shown to be a form of electromagnetic radiation. Doctors now use X-rays to produce internal photographs of the body.

AN ELECTROMAGNETIC WAVE

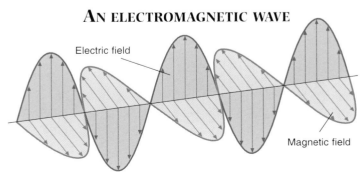

Electric field

Magnetic field

In electromagnetic waves, electric and magnetic fields travel along at right angles to each other.

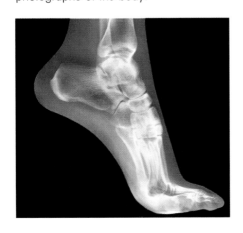

RADIO WAVES

In the 1880s, German physicist Heinrich Hertz put Maxwell's electromagnetic theories to the test. Maxwell had suggested that electromagnetic waves could be produced by setting up a vibrating electric charge. So Hertz used a special electric circuit to do just that. The electrical effect on another, unconnected circuit proved both that waves had been generated, and that they had travelled through the air. What is more, they moved at the speed of light, but had a longer wavelength. As Maxwell had predicted, a new form of electromagnetic wave had been discovered. These waves are now known as radio waves.

James Clerk Maxwell

Clarendon Press Series

A TREATISE

ON

ELECTRICITY AND MAGNETISM

BY

JAMES CLERK MAXWELL, M.A.

LLD. EDIN., F.R.SS. LONDON AND EDINBURGH
HONORARY FELLOW OF TRINITY COLLEGE,
AND PROFESSOR OF EXPERIMENTAL PHYSICS
IN THE UNIVERSITY OF CAMBRIDGE

VOL. I

Oxford

AT THE CLARENDON PRESS
1873

[All rights reserved]

The title page of *A Treatise on Electricity and Magnetism*, Maxwell's second major work, published in 1873.

Maxwell sent his 1856 paper on electromagnetism to Michael Faraday. The following extract is taken from Faraday's reply.

Albemarle Street, London is still the headquarters of the Royal Institution, where Faraday worked (see pages 28-9).

This means 'it is work that I am grateful for'.

Albemarle Street, W.,
25th March 1857

MY DEAR SIR—I received your paper, and thank you very much for it. I do not say I venture to thank you for what you have said about "Lines of Force," because I know you have done it for the interests of philosophical truth; but you must suppose it is work grateful to me, and gives me much encouragement to think on. I was at first almost frightened when I saw such mathematical force made to bear upon the subject, and then wondered to see that the subject stood it so well...

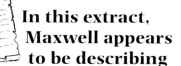
In this extract, Maxwell appears to be describing bells in a belfry. But in fact he is talking about electricity and magnetism. He explains that these forces may appear to act independently but they are, in fact, inseparable.

In an ordinary belfry, each bell has a rope which comes down through a hole in the floor to the bellringer's room. But suppose that each rope, instead of acting on one bell, contributes to the motion of many pieces of machinery, and that the motion of each piece is determined not by the motion of one rope alone, but by that of several, and suppose, further, that all this machinery is silent and utterly unknown to the men at the ropes, who can only see as far as the holes in the floor above them.

ERNEST RUTHERFORD
AND SUBATOMIC PARTICLES

John Dalton (see pages 18-19) believed atoms were indivisible particles. But from the late 19th century, this theory was gradually demolished by a number of physicists, including the great Ernest Rutherford (1871-1937).

By the mid 19th century, many scientists were using the new technique of spectroscopy (see page 8) to analyse gases. To do so, they had to make the gases glow so that the light from them could be examined. They achieved this effect by putting the gases one by one in a special glass tube, then passing electricity through them. The first unexpected result of such experiments was the discovery of X-rays in 1895 (see page 30). Further intriguing developments soon followed.

Joseph John Thomson, Professor of Experimental Physics at Cambridge University, investigated another type of ray that appeared in the gas tests. These rays emerged from the cathode in the glass tube, so were called cathode rays. By 1897, Thomson had proved cathode rays were made of negatively charged particles. Later, he calculated that each particle's mass was far less than the mass of any atom. The tiny new particles were christened electrons. Thomson suggested they were stuck in the positively charged bodies of atoms like plums in a pudding.

New Zealander Ernest Rutherford, a former pupil of Thomson's, shed more light on atomic structure. His investigations of radioactivity showed that there were at least two types, which he called alpha and beta particles, and that both were emitted by atoms as they spontaneously disintegrated. In 1910, while a professor at Manchester University, Rutherford instructed two of his assistants to fire alpha particles at gold foil. The results were astonishing. Most particles went straight through the foil, suggesting that the atoms in it were made up of empty space and of electrons with little mass that let particles through easily. However about 1 particle in 8000 was greatly deflected. This suggested that the atoms in the foil also had tiny, dense cores which caused particles that hit them to bounce back. In 1911, Rutherford proposed an atomic structure based on this idea – a positively charged centre, now called the nucleus, with negatively charged electrons in orbit around it. In 1913, the Danish physicist Niels Bohr devised an improved version of this model.

When Rutherford was back at Cambridge, in 1919, he carried out another major experiment, bombarding nitrogen with alpha particles. This caused the gas's atomic nuclei to disintegrate and to emit positively charged particles that Rutherford called protons. Protons were shown to be present in the nuclei of all atoms, lending support to Rutherford's model of atomic structure. In 1932, British physicist James Chadwick showed there was another type of particle in the nuclei of all elements except hydrogen. This was the neutron, which has no electric charge.

These ideas of atomic structure allowed scientists to explain many of the properties of elements. But there was much more to discover.

Ernest Rutherford (facing out of the picture) in the Cavendish Laboratory at Cambridge University.

PILES OF PARTICLES

Since the 1950s, machines called particle accelerators have been used to break up atomic nuclei. In this way, scientists have discovered many new types of subatomic particle. They now know of over 300, which can be divided into two groups: elementary particles and hadrons. Elementary particles, which include quarks and leptons, cannot be split any further. The electron is a type of lepton. Hadrons can be split into quarks. Protons and neutrons are types of hadron. One of the most important scientists to investigate elementary particles was American Murray Gell-Mann. He was the first to suggest the existence of quarks, in 1964, and gave them their unusual name.

INSIDE ATOMS

Rutherford model of an atom

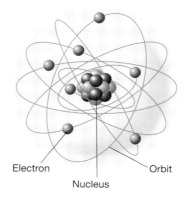

Electron — Orbit
Nucleus

Revised Rutherford-Bohr model of an atom

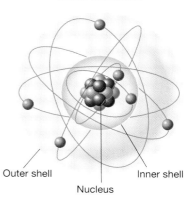

Outer shell — Inner shell
Nucleus

The diagram on the left shows an atom with electrons moving in orbits around the nucleus, as Rutherford proposed. To explain why the electrons do not spiral into the nucleus, Niels Bohr proposed that the electrons could only move in shells (as shown on the right). Today, scientists believe that electrons do not move along fixed paths within their shells, but can be found in broad regions called 'orbits'.

Rutherford probably wrote this page in his notebook during the winter of 1910-11. Headed 'Theory of structure of atom', it shows the first, rough diagram he made of his new atomic model.

The Tevatron particle accelerator at Fermilab, near Chicago, USA. Particle accelerators are huge machines that hurl elementary particles along at vast speeds, often causing them to split and produce new particle types. This particle accelerator is the largest in the world.

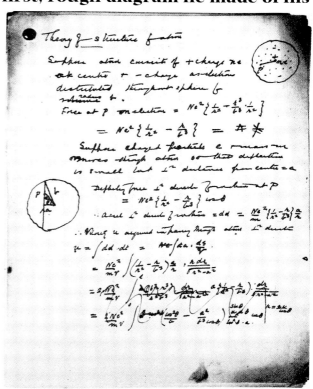

Rutherford's diagram is accompanied by equations and notes that provide support for his new theory of the atom.

ALBERT EINSTEIN
AND RELATIVITY

The German-born physicist Albert Einstein was the greatest scientist of the 20th century. But his famous theories of relativity, and many of his other ideas, can be extremely difficult for non-specialists to grasp.

Einstein did not excel in his early years at school (see box). Nevertheless, he continued with his studies and in 1905 received a science doctorate from the University of Zurich. In the same year he published four brilliant papers in a journal called *Annalen der Physik* (*Annals of Physics*) and his genius became clear. One of these papers, *On the Electrodynamics of Moving Bodies*, introduced the Special Theory of Relativity. It was Einstein's solution to a problem that had long been exercising his mind. This problem was an apparent conflict between two major theories.

The first theory, formulated by Isaac Newton in the 17th century, said that all speeds are relative. For example, if a car is travelling at 80 km/h, it is moving at exactly that speed relative to a stationary observer. But relative to another car travelling

ALBERT EINSTEIN (1879-1955)

Albert Einstein (see picture below) was born into a Jewish family in Ulm, Germany. In 1880 the family moved to the city of Munich. Here Einstein began his schooling, but his progress was slow compared with other pupils. As he moved into his teenage years, Einstein found the rigid school structure increasingly irksome and disrupted his lessons. In 1894, his parents moved once more, this time to Milan in Italy. Einstein stayed behind, but was expelled from school. He eventually finished his secondary education at Aarau in Switzerland and went on to the Federal Institute of Technology in Zurich. After graduating, Einstein was not given a post at the institute as he had hoped, so became a private tutor. In 1902, he got a job at the Swiss Patent Office in Bern, at the same time continuing to work for a PhD at the University of Zurich and to think about physics problems.

The year 1905, when he not only gained his doctorate but published four major scientific papers (see main text), was a turning-point in Einstein's life. His work swiftly brought him recognition – and job offers. Eventually, in 1914, he settled at the Kaiser Wilhelm Institute for Physics in Berlin. Einstein remained in Germany until 1933, when the Nazis, who hated and persecuted Jews, came to power. They forced him to leave his job, so he moved to the USA, where he became Professor of Mathematics at Princeton University. During the later years of his life, Einstein combined science with campaigns for pacifism (see box page 36) and for Zionism, the movement for the creation of a Jewish state in Palestine.

towards the first one at 60 km/h, it is approaching at 140 km/h. The second theory, formulated by James Clerk Maxwell (see page 30), was that the speed of light never changes. The speed at which the light source or the person observing the light is moving makes no difference.

In his Special Theory, Einstein stated that Maxwell was correct. Newton's method of calculating relative speeds would, he predicted, break down as objects approached the speed of light. Nothing, Einstein believed, could travel faster than the speed of light. So the combined speeds of two objects moving towards each other could never exceed it, no matter how fast they were going. The closer they got to the speed of light, the more meaningless the calculation would become.

Einstein's theory also states that as an object approaches the speed of light, its length decreases while its mass increases. A clock on board will also go more slowly the faster the object travels, a phenomenon called time dilation. These effects happen because measurements of space and time are relative, not absolute as once believed. They change according to the context, like speeds. Scientists have proved these ideas by sending tiny particles around particle accelerators (see page 33) at great rates. ➲

English physicist Isaac Newton (1642-1727) deduced the existence of gravity and devised three laws of motion that explained how forces act on objects to produce or prevent movement. Until Albert Einstein introduced the concept of relativity, Newton's ideas were considered to provide a complete picture of the forces at work in the Universe.

SEEING THE LIGHT

A second paper published by Einstein in 1905 (see main text) contained another idea just as revolutionary as the theory of relativity. At this time, scientists were puzzled by the discovery that when light is directed at some metals, they emit electrons. Even top physicists could not understand how light waves could produce this photoelectric effect. In his paper, Einstein showed that light existed and acted as particles, as well as waves, and that the particles' special properties accounted for the effect. He called these particles quanta, but they are now known as photons. This new understanding of light helped to transform physics.

Even Einstein found his own theories difficult to comprehend at first! This is what he said about the period of his life when he was formulating the Special Theory of Relativity.

I must confess that at the very beginning when the Special Theory of Relativity began to germinate in me, I was visited by all sorts of nervous conflicts. When young I used to go away for weeks in a state of confusion, as one who at that time had yet to overcome the state of stupefaction in his first encounter with such questions.

Stupefaction means 'astonishment and confusion'.

The Special Theory applied only to objects that were stationary or moving at fixed speeds. After its publication, Einstein worked on an extension that would deal with accelerating objects, too. This was published as *The Foundation of the General Theory of Relativity* in 1916. The General Theory is a complex piece of mathematics. But one of its central ideas, about gravity, can be explained more simply. Isaac Newton had taught that gravity was a force operating between all bodies with mass, pulling them towards one another. The more massive the object, the greater the force it exerts. Einstein stated that gravity was not a property of mass, but of space. Massive bodies distort space so that it is curved, rather as if you dropped a ball into a flat rubber sheet (see diagram page 37). Less massive bodies, as well as light rays, 'fall' into the apparent hollow.

Despite his brilliance, Einstein did make some errors. Using his own General Theory equations, he calculated that the Universe was not expanding but static. He later described this conclusion, corrected during his own lifetime by Edwin Hubble and others (see pages 10-11), as 'my biggest mistake'. Einstein also stumbled over quantum theory, which was developed by German physicist Max Planck in 1900. It states that matter emits and absorbs energy in tiny units called quanta. Einstein himself successfully applied the theory to light (see box page 35). However, later developments led him to believe that it was flawed. The theory has since been generally accepted.

These lapses did little to dent Einstein's reputation, which had grown fast following the publication of the General Theory. His worldwide fame was assured after the 1919 solar eclipse (see document page 37) proved the theory was correct. Einstein's insights, although they remained a mystery to most people, ensured that science was never the same again.

EINSTEIN AND THE BOMB

Einstein's fourth paper of 1905 contained the famous equation $E = mc^2$, which shows the relationship between energy (E) and mass (m). The amount of energy locked up in a piece of matter is equal to its mass multiplied by the speed of light squared (c^2). As the speed of light is very fast, a mass can, potentially, be converted into a huge amount of energy. When Einstein wrote this equation, it was pure science. But as the Second World War approached, the Nazis began to use the principle to develop the atomic bomb. In 1939, Einstein and other scientists wrote to the American president, Franklin D. Roosevelt, urging that the USA carry out similar research. However, when the Americans finally dropped an atomic bomb on Hiroshima, Japan in 1945, Einstein was appalled by the devastation. He became a committed pacifist.

The light-bending effect that proves Einstein's Theory of Relativity can be measured during a total solar eclipse. This eclipse was seen in many parts of the world on 11 August 1999.

Hiroshima, Japan, after the atomic bomb was dropped. By the end of 1945, about 140,000 people had died as a result of the bomb's terrible power.

GRAVITY AT WORK

Einstein explained how gravity works by showing that massive objects cause the space around them to curve (see text page 36). As a result, less massive objects 'fall' towards them.

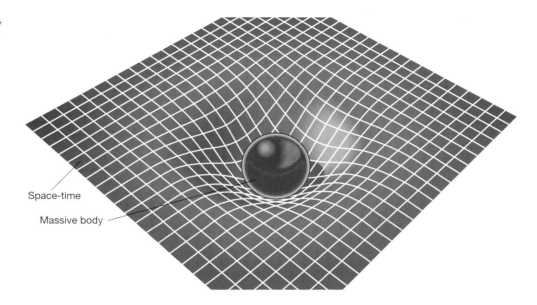

Space-time

Massive body

 Einstein worked out the idea that massive objects in space cause light to bend some years before the General Theory of Relativity was published. In this letter of 1913, written in German to American astronomer George Ellery Hale, Einstein drew a diagram to show the bending effect. This effect was proved when there was a total eclipse of the Sun in 1919. The event created conditions in which British astronomer Arthur Eddington was able to test whether the path of starlight was affected by the Sun's massive bulk. It was, exactly as Einstein had predicted.

0.84" is the number of seconds (") by which Einstein predicts the starlight will be deflected. A 'second' here is a subdivision of an angle.

Stern is German for 'star'.

This line shows the expected path of the starlight.

Sonne is German for 'Sun'.

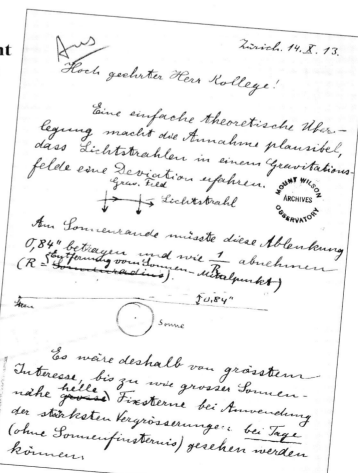

CHAPTER 4

GEOLOGISTS

CHARLES LYELL
AND UNIFORMITARIANISM

The holy books of many religions relate when and how the world was created. These ideas went unchallenged for thousands of years. Then some people began to take a more scientific approach, investigating the Earth's surface for evidence. By the 19th century, geology was an accepted science, and Charles Lyell (1797-1875) one of its most important figures.

Many early geologists were miners seeking coal and iron to supply the factories of the Industrial Revolution. In the 18th century, Abraham Werner, a teacher at a German mining academy, proposed a geological theory. A flood, he said, had once covered the Earth and the planet's rock strata were formed from minerals in the water that sank to the seabed or crystallised as the ocean evaporated.

The Scottish geologist James Hutton disagreed with Werner's ideas. In 1795, he put forward his own views in a book, *Theory of the Earth*. The Earth's surface, Hutton said, had not been shaped by floods or other sudden catastrophes. It had developed, and was still developing, as a result of processes such as erosion, sedimentation and heating. This heating occurred inside the Earth. It made volcanoes erupt and pushed rocks up to create folds and mountains. New rocks formed as volcanic lava and molten material under the surface solidified.

Hutton's theory, which he called uniformitarianism, won support from other geologists. They included his fellow Scot, Charles Lyell. As a young law student with a keen interest in geology, Lyell had travelled widely in England, France and other parts of mainland Europe during the early 19th century. On his trips, he had studied the landscape, examining any fossils he found there. These thorough investigations had convinced him that Hutton's ideas were correct.

Eventually, Lyell decided to write a book in support of Hutton's theory. This three-volume work was called *The Principles of Geology* (1830-3). In it, Lyell explained how physical processes

In this cartoon, James Hutton chips away at rocks in the shapes of the faces of his scientific enemies.

had created the landscape. He also used fossil evidence to date the events that he described. The work was an instant hit. Among its readers was naturalist Charles Darwin (see pages 46-9), who later used some of Lyell's ideas.

During the course of his life, Lyell made many more contributions to the study of the Earth, in particular developing an early form of geological time scale, as well as ensuring that uniformitarianism was widely accepted. By the late 19th century, geology had become a strong, evidence-based science with a sound framework on which future scientists could build.

Charles Lyell

Lyell formulated his theories by studying landscape. He was particularly intrigued by the volcanic plugs – lumps of solidified magma in the tops of extinct volcanoes – in central France.

TELLING THE TIME

William Smith was a 19th-century British surveyor whose job involved examining land to build canals. As he worked, Smith noticed animal fossils in the rock strata. Then he realised that certain types of fossil occurred in the lower, older layers of rock and other, different types in the younger layers on top. In other words, fossil types were a guide to the age of rocks. This insight was a great help to early geologists, but during the 20th century far more accurate dating techniques were devised (see page 42).

The first of these two extracts is taken from a letter Lyell wrote just before the publication of *The Principles of Geology*. The second comes from a letter written during a trip to the USA. It discusses one of the fossils examined by Lyell and outlines the problems endured by a geologist whose ideas conflicted with Biblical teaching.

These few lines are a brief description of the theory of uniformitarianism.

Abstract is another word for 'summary'.

Naples: January 15, 1829.
...My work is in part written, and all planned. It will not pretend to give even an abstract of all that is known in geology, but it will endeavour to establish the principle of reasoning in the science; and all my geology will come in as illustration of my views of those principles... which, as you know, are neither more nor less than that no causes whatever have from the earliest time to which we can look back, to the present, ever acted, but those now acting; and that they never acted with different degrees of energy from that which they now exert.

The **Cheirotherium** was a type of ancient, four-legged reptile.

In the 17th century, a Christian archbishop called James Ussher had used Bible texts and the works of ancient writers to calculate that the world was created in 4004BC.

Philadelphia: April 27, 1846
I had some thoughts before my arrival here, to draw up a paper on what I learnt and observed... in the Pennsylvanian Coal Field... I have satisfied myself that Dr. King is right in believing that he has discovered in the middle of the coal formation, the foot tracks of a large reptilian quadruped or animal allied to the Cheirotherium... Dr. King was a beginner in geology when he first found, and to his credit appreciated duly, the importance of the Cheirotherium tracks. He is a man of thirty years of age, and in an extensive medical practice, who has suffered some persecution, professionally and socially, for believing the world to be more than 6,000 years old...

ALFRED WEGENER
AND CONTINENTAL DRIFT

In the 19th century, geologists discovered how the Earth's rocks and natural features formed and changed (see pages 38-9). In the 20th century, a few of them, including the German Alfred Lothar Wegener, began to investigate a puzzle on a larger scale. They suspected that the world's great landmasses, the continents, had moved over time. But they needed evidence to prove their theory to fellow scientists.

Anyone looking at a world map can see that many separate pieces of land would fit neatly into one another if they were pushed together across the sea. For example, the point on the east coast of South America is just the right shape to fit underneath the bulge on the west coast of Africa. In the 19th century, some geologists began to ponder what this might reveal about the history of the Earth.

The Austrian scientist Eduard Suess, Professor of Geology at Vienna University from 1861, was the first to consider the question in detail. He discovered that rocks on coasts that faced one another across the sea often matched exactly. He also found that many lands now separated by

ALFRED LOTHAR WEGENER (1880-1930)

Alfred Lothar Wegener (see picture below) was born in Berlin, Germany in 1880. He studied astronomy at universities in both Germany and Austria, but his other great interest was meteorology, and in 1908 he became a lecturer in both subjects at the German university of Marburg. It was while there, in 1915, that he wrote his famous book *Die Entstehung der Kontinente und Ozeane* (see main text). Four years later, after the First World War, Wegener joined the meteorological department of the German Marine Observatory and began making trips to Greenland to carry out research. In 1924, he was appointed Professor of Meteorology and Geophysics at Graz University in Austria. Wegener died during his fourth trip to Greenland.

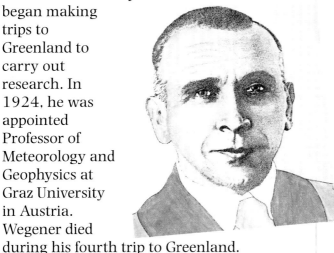

CONTINENTAL DRIFT

These accurate, modern continental drift maps show the Earth's landmasses as they were 180 million years ago, 120 million years ago and today.

ocean contained fossils of the same plants and animals (see box page 43). All this evidence led Suess to claim that Australia, India, Africa and South America had originally formed one vast southern continent. In the 1890s, he christened this landmass Gondwanaland.

In the early 20th century, an even more startling theory about the continents was proposed by German scientist Alfred Wegener. His 1915 book *Die Entstehung der Kontinente und Ozeane*

(*The Origin of Continents and Oceans*) made the bold suggestion that all the world's continents had once been a single giant supercontinent. He called this landmass Pangaea, Greek for 'All Earth'. Like Suess, Wegener used rock and fossil evidence to support his idea. He also pointed out that the continental shelves, the underwater ledges that surround continents, would fit together even better than the coastlines. They matched like the pieces of a jigsaw. But Wegener's most daring step was to ➲

Alfred Wegener rewrote parts of *Die Entstehung der Kontinente und Ozeane* (1915) several times in order to add information provided by Alexander Du Toit (see page 42) and others. The English translations, from which this and the extract on page 43 are taken, included all these new facts. In the extract below, Wegener is explaining how rocks and their formations can provide evidence of continental drift theory.

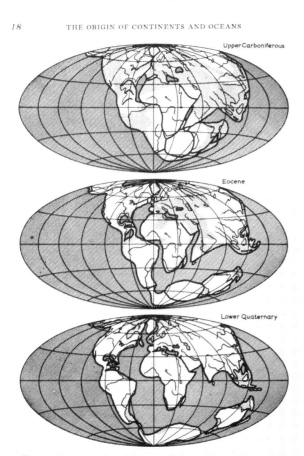

18   THE ORIGIN OF CONTINENTS AND OCEANS

Upper Carboniferous

Eocene

Lower Quaternary

FIG. 4.   Reconstruction of the map of the world according to drift theory for three epochs.
Hatching denotes oceans, dotted areas are shallow seas; present-day outlines and rivers are given simply to aid identification.   The map grid is arbitrary (present-day Africa as reference area; see Chapter 8).

These three maps from *Die Entstehung der Kontinente und Ozeane* show how Wegener believed the world's continents had begun as a single landmass (top), then gradually drifted apart.

A **rift** is a giant crack in the Earth's crust.

By comparing the geological structure of both sides of the Atlantic, we can provide a very clear-cut test of our theory that this ocean region is an enormously widened rift whose edges were once directly connected, or so nearly as makes no difference... alkali-rich rock is... to be found on exactly corresponding stretches of coastline: on the Brazilian side, at various places in the Serra do Mar... on the African side, on the coast of Lüderitzland, at Cape Cross north of Swakopmund, but also in Angola... both in Brazil and South Africa... beds yield the famous diamond finds. In both these regions the peculiar type of stratification known as "pipes" occurs. There are white diamonds in Brazil in Minas Geraes State and in South Africa north of the Orange River...

Stratification means 'rock layering'.

propose a theory about how the continents had changed position.

Wegener called his theory continental drift. It suggested that Pangaea broke up about 200 million years ago and that the resulting pieces slowly drifted apart. On research trips to Greenland, Wegener had taken measurements that showed this drifting was still taking place. The measurements themselves were very inaccurate, and have since been revised using lasers, but the fact of movement was clear. Many professional geologists ridiculed Wegener's idea. Their main objection was that the continents, which are largely made of a rock called granite, could not have moved through the solid basalt rock below.

After his book was published, Wegener continued his academic career. Meanwhile, two other geologists were carrying out research that supported his ideas. The first was South African Alexander Du Toit. While working for the Geological Commission of the Cape of Good Hope, Du Toit studied South African rocks in great detail. In 1923, he carried out similar work in South America. His results convinced him the two continents had once been joined. Du Toit outlined his ideas in *A Geological Comparison of South America with South Africa* (1927).

Wegener's second supporter was British geologist Arthur Holmes. His main interest was rock-dating techniques (see box). But in 1929, he proposed an idea that helped explain continental drift, too. Holmes' idea was that heat generated inside the Earth caused rocks below the surface to circulate in convection currents. As the rocks rose, Holmes thought, they would push up and out with great force, driving continents apart.

Despite these developments most geologists continued to reject the idea of continental drift. Its mechanism was not understood until years later (see pages 44-5).

ROCKS AND RADIOACTIVITY

Using fossils to date rock layers (see page 39) was a major breakthrough for geologists. But this method could reveal only the relative ages of rocks, that is if a particular rock was older or younger than another. The absolute age of rocks, that is how many years they had existed, remained a mystery. Then, in the 20th century, scientists discovered that the atomic nuclei of radioactive materials emit particles at a fixed rate. These particles are often trapped inside radioactive rocks. Once their emission rate is known, scientists can determine the age of a rock by calculating how many particles it contains. Arthur Holmes (see main text) used radioactive techniques to devise the first geological time scale that divided the Earth's history into dated periods. It appeared in 1913.

Arthur Holmes

Ernest Rutherford (see pages 32-3) made the following statement in 1907. It indicates he was the first person to realise that the rate of radioactive decay could be used to date rocks (see box).

Alpha particles, a type of radioactivity, are chemically the same as atomic nuclei of the element helium.

If the rate of production of helium from known weights of the different radio-elements were experimentally known, it would thus be possible to determine the interval required for the production of the amount of helium observed in radioactive minerals, or, in other words, to determine the age of the mineral.

FOSSIL EVIDENCE

The location of plant and animal fossils provided important evidence for continental drift theory. Fossils of the *Glossopteris* fern were found in Antarctica, Australia, India, Africa and South America, supporting the idea of an ancient southern supercontinent. Fossils of an extinct reptile called *Mesosaurus* were discovered in South Africa, Brazil, Paraguay and Uruguay, again indicating that Africa and South America were once joined. The former link between Europe and North America was suggested by the presence of *Diadectid* insect fossils in both places.

A ring-tailed lemur from Madagascar, an island off the east coast of Africa. All 21 species of lemur are native only to that island. This is because Madagascar split from Africa about 60 million years ago, so the animals there evolved separately.

The location of living plants and animals, as well as of fossils, can help geologists to work out where continents joined and when they separated. In this second extract from his book (see page 41), Wegener explains what Australia's animals reveal about the Earth's past. He has already stated that they can be divided into three groups, and that each group has lived in the country for a different length of time.

Fauna means 'animal life'.

Correlation means 'relationship' or 'connection'.

Marsupials are mammals whose young are born in an undeveloped state and continue to grow in a pouch outside their mother's body. They include kangaroos, koalas and opossums.

The **Sundas** are a chain of islands in the Pacific and Indian oceans. They include Java and Borneo.

The most ancient element, generally encountered in southwestern Australia, shows some interrelationship with the fauna of India... Here warmth-loving animals are representative of the relationship, and so are the earthworms... The correlation dates back to the time when Australia was still joined to India. The second Australian faunal element... includes those peculiar mammals, the marsupials... This element shows points of relationship with South American fauna. Apart from Australia... marsupials now live mainly in South America... The third Australian faunal element is the most recent, and emigrated from the Sundas; it has... already taken over the northeast of Australia. The dingo (wild dog), rodents, bats and other mammals immigrated to Australia in the post-Pleistocene... This threefold subdivision... agrees most elegantly with drift theory. One need only look at the three reconstruction maps [see page 41] and the explanation will be immediately forthcoming.

The **Pleistocene** is a geological era that began about 1.8 million years ago and ended about 10,000 years ago.

HARRY HESS
AND SEA-FLOOR SPREADING

Geologists eventually solved the puzzle of continental drift (see page 42) by examining the Earth's ocean floors. American geologist Harry Hammond Hess played a major role in this investigative work.

Exploration of the world's ocean floors began in earnest during the 19th century. Gradually, using sonar and other techniques, scientists discovered that the floors were not completely flat but made up of mountains, trenches and other natural features. Then, in the 1950s, American oceanographer William Maurice Ewing showed that a 65,000-km long mountain range snaked right around the world in the middle of the ocean floor. In 1957, he also demonstrated that giant rifts (cracks) ran all along the centre of this mid-ocean ridge.

Harry Hess, Professor of Geology at Princeton University in the USA, followed these developments closely. He had read Wegener and Du Toit's works (see pages 40-3), and was intrigued by the continental drift theory. However, he doubted its truth as he knew continents did not have the in-built energy to move across the Earth. But then Ewing's findings, and his own discovery that ocean floor rocks are younger than the rocks of the surrounding continents, gave him a revolutionary idea.

In 1962, Hess revealed his idea when he published a paper called *History of Ocean Basins*. It was a description of the way he believed the ocean floor had formed. Molten rock called magma, Hess suggested, was always rising from inside the Earth and oozing through the rifts in the mid-ocean ridge. As it did so, it pushed existing rock away from the ridge, then solidified to form new, younger rock. Hess's theory became known as seafloor spreading.

British geologists Fred Vine and Drummond Matthews set out to test Hess's idea. They knew that the Earth's magnetic field periodically reverses itself, the north magnetic pole becoming the south, and the south becoming the north. They also knew that, when magma cools, evidence of the field's direction at the time is preserved in the solid rock that forms. So, gradually, series of rock 'stripes' with opposing magnetic fields are created. If Hess was right, the magma from the mid-ocean rifts would have formed stripes in the surrounding ridge.

The two men found that the ridge was indeed made up of rock stripes, which could be dated. In addition, the stripes were symmetrical on the two sides of each central rift, indicating that both had formed from magma leaving the rift at the same time. Vine and Matthews published their findings

SEA-FLOOR SPREADING

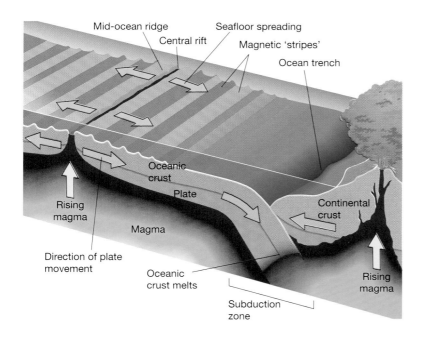

Mid-ocean ridge
Central rift
Seafloor spreading
Magnetic 'stripes'
Ocean trench
Oceanic crust
Plate
Rising magma
Magma
Continental crust
Direction of plate movement
Oceanic crust melts
Rising magma
Subduction zone

As magma rises through rifts in the sea floor, it forces the surrounding rock apart. As it solidifies, it creates new rock.

in 1963, and seafloor spreading was accepted. Scientists soon realised that Hess's theory helped to explain continental drift. The continents did not, in fact, drift apart. They were pushed apart by rock emerging from the ocean floor. But there was still more to discover (see box).

HARRY HAMMOND HESS (1906-69)

Harry Hammond Hess (see picture below) was born in New York City, USA. After graduating from Yale University in 1927, he worked as a geologist in Northern Rhodesia. In 1934, Hess joined the staff at Princeton University. At the same time as pursuing an academic career, Hess served in the US Navy. While surveying the Pacific Ocean floor, he discovered many flat-topped mountains, previously unknown features that he christened guyots. The survey work prompted Hess's interest in submarine geology and led him to develop the seafloor spreading theory (see main text).

PLATE TECTONICS

New ideas about the Earth's structure helped to improve geologists' understanding of seafloor spreading. The Earth's crust – the rock that makes up the ocean floor and the continents – forms only the top part of a deeper layer. In the 1960s, Canadian geologist John Tuzo Wilson and others showed that this layer is divided into about 15 giant pieces, called 'plates'. The rifts in the ocean floor are plate edges. As new rock is formed there it forces the plates apart and the plates carry the continents with them. The opposite process occurs where plate edges collide and one forces the other below the surface, known as subduction zones. The study of plate movement is known as plate tectonics.

The San Andreas Fault, California. It is a transform fault, produced when two plates slide past each other.

Hess was delighted that his seafloor spreading theory caused such a flurry of interest. But he was alarmed that so much of the following research was not carefully planned. In the introduction to a paper he published in 1965, he included this warning.

A **hypothesis** is an untested scientific theory. Hess is saying that it is pointless for scientists to carry out research and collect information if they have no idea what they are trying to prove or disprove. A hypothesis, even if it is wrong, still provides the necessary framework for worthwhile scientific research.

To bring the problem into focus and guide continued exploration, co-ordinating hypotheses are needed and necessary. Even incorrect or partly incorrect speculations serve to identify the crucial observations needed for progress. Blind... collection of data without a framework of hypothesis by which it can be tested is wasteful and commonly unproductive, and leads to an accumulation of an indigestible mass of data of minor significance.

CHAPTER 5

BIOLOGISTS
CHARLES DARWIN
AND EVOLUTION

As a young man, British naturalist Charles Robert Darwin briefly trained for the Anglican priesthood. Eventually, however, he abandoned his religious studies for a scientific career. The results were profound – not only a revolution in biology, but turmoil in the Church.

In 1831, following his graduation from Cambridge University (see box), Darwin was appointed as the naturalist on board HMS *Beagle*, a ship heading for South America and the Pacific Ocean to carry out coastal surveys. The *Beagle* remained at sea for five years. During that time, Darwin explored many places, marvelling at the variety of animals and plants, and collecting specimens.

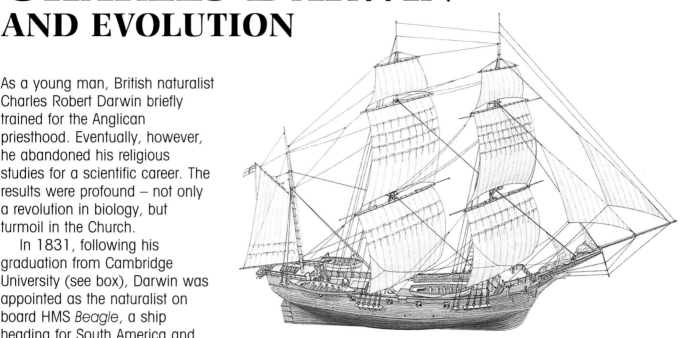

HMS *Beagle*, the ship on which Darwin made his momentous journey. It carried small boats that he and other passengers used for exploring small islands and rivers.

He also examined landscapes, using *The Principles of Geology* by Charles Lyell (see page 38) as his guide.

In 1839, three years after his return to England, Darwin published an account of his trip, *A Naturalist's Voyage on the* Beagle.

CHARLES DARWIN (1809-82)

Charles Darwin (see picture right) was born in Shrewsbury, England. He enrolled to study medicine at Edinburgh University in 1825, but soon realised he had made a mistake. The lectures bored him and the demonstrations on patients were an ordeal. So he left for Cambridge University, where he prepared for a life as a priest in the Church of England. Soon, however, he grew far more interested in geology and botany. The Professor of Botany, John Henslow, became a friend, and in 1831 recommended Darwin for a job on the *Beagle*. Darwin combined the eventful career that followed (see main text) with a full personal life – he married his cousin and they had ten children. Eventually, he became ill with Chagas disease, which was caused by a parasite that he had picked up during his voyage to South America. He died in 1882 and was buried in Westminster Abbey.

But he continued to ponder the deeper significance of what he had seen. Darwin had been astounded by the sheer number of animal species, and by the fact that they often differed only in small details. For example, on the Galapagos Islands he had counted 14 types of finch. They were all similar, but their beaks varied according to the type of food they ate. In the same way, Galapagos tortoises had necks of different lengths, depending on whether they ate leaves from trees or ground plants.

All this evidence led Darwin to develop a new theory. He began to doubt the Christian teaching that God had created each animal species, which never altered. Instead he started to think that each species had evolved, that is changed over time, so that it could survive. In the case of the Galapagos finches he suggested that a single finch species had arrived on the islands from mainland South America. It had then evolved into many species, each adapted to conditions and food sources in the particular area where it settled. ➲

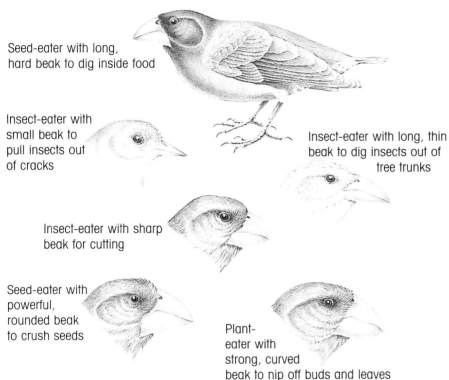

Six of the 14 kinds of finch that Darwin found on the Galapagos Islands. Each one has a different type of beak ideally suited to the type of food it eats.

Seed-eater with long, hard beak to dig inside food

Insect-eater with small beak to pull insects out of cracks

Insect-eater with long, thin beak to dig insects out of tree trunks

Insect-eater with sharp beak for cutting

Seed-eater with powerful, rounded beak to crush seeds

Plant-eater with strong, curved beak to nip off buds and leaves

The following extract comes from Chapter Three of *The Origin of Species* (see page 48), called 'The Struggle for Existence'. The words provide a vivid description of this struggle, which leads to the 'survival of the fittest'.

People of the early cultures that flourished in south-eastern America, for example the Adena, Hopewell and Mississippians, built mounds that were often used for burials or religious rituals.

When we look at the plants and bushes clothing an entangled bank, we are tempted to attribute their proportional numbers and kinds to what we call chance. How false a view this is!... [It] has been observed that the trees now growing on the ancient Indian mounds, in the Southern United States, display the same beautiful diversity and proportion of kinds as in the surrounding virgin forests. What a struggle between the several kinds of trees must here have gone on during long centuries, each annually scattering its seeds by the thousand; what war between insect and insect – between insects, snails, and other animals with birds and beasts of prey – all striving to increase, and all feeding on each other or on the trees or their seeds and seedlings, or on the other plants which first clothed the ground and thus checked the growth of the trees!

Now Darwin began to consider what the mechanism for evolution might be, that is how species changed. The answer came to him after he read an essay by economist Thomas Malthus. It said that when human populations could not produce enough food for all, the weakest people starved, died of disease or were killed in fighting. In other words, only the strong survived. Darwin saw that this could apply to animals, too.

Darwin set about working out the details. He realised that when animals were competing for food, or struggling to live in a harsh environment, their physical characteristics affected their ability to survive. For example, birds with strong beaks can crack open seeds easily, and animals with powerful legs can escape their predators. So, Darwin reasoned, creatures with useful characteristics lived long enough to breed and pass them on to offspring. By contrast, animals without them died out. Darwin called this process natural selection.

Darwin agonised for many years about revealing his theories to the public, because he foresaw the uproar they would cause. But eventually, in 1859, he presented them in a book, *The Origin of Species by Means of Natural Selection*. It became a best-seller, but provoked fury among Christians who believed in the literal truth of the Bible. The idea that natural selection could lead to the changing of one species into another caused particular anger. This was because it implied humans were not created directly by God but had evolved like other creatures. In fact the book only hinted at this conclusion, but the controversy still raged (see box below). Despite the hostile reaction, Darwin continued his research and published several more books. They included *The Descent of Man* (1871), which discussed human evolution in some detail. Darwin's works are still widely read today, and his ideas are generally accepted by the scientific community. However, some Christians continue to reject the theory of evolution and to claim that the world and everything in it were created by God, exactly as related in the Bible.

ALFRED RUSSEL WALLACE

Darwin was not the only person to formulate the theories of natural selection and evolution – Welshman Alfred Russel Wallace reached similar conclusions in a similar way. In the first half of the 19th century, Wallace travelled widely across South America and the Malay Archipelago. Like Darwin he collected specimens, wrote books and thought deeply about how species might have developed. In 1858, Wallace wrote a paper called *On the Tendency of Varieties to Depart Indefinitely from the Original Type* and sent it to Darwin for comment. Darwin realised Wallace had the same ideas as him, so quickly finished *The Origin of Species* before Wallace had the chance to publish first.

'DARWIN'S BULLDOG'

Darwin knew *The Origin of Species* would cause great upset among religious people, as well as some scientists. But he was a naturally retiring man who had no wish to join in the public debate. Fortunately, he had many able defenders who were willing to do just that. Chief among them was British biologist Thomas Henry Huxley (see picture below). In 1860, Huxley faced two of Darwin's most outspoken opponents at a specially arranged meeting in Oxford. They were the Bishop of Winchester, Samuel Wilberforce, and a fossil expert called Richard Owen. Despite his reputation as a smooth speaker, 'Soapy Sam' Wilberforce was outwitted by Huxley's firm grasp of the arguments and refusal to be angered by taunts. As a result of this and many other displays of debating skill, Huxley became known as 'Darwin's bulldog'.

This extract comes from Chapter Four of *The Origin of Species*, called 'Natural Selection'. In it, Darwin explains how this process has caused some insects and birds to develop camouflage colourings, making them difficult for predators to see and catch.

Although natural selection can act only through and for the good of each being, yet characters and structures, which we are apt to consider of trifling importance, may thus be acted on. When we see leaf-eating insects green, and bark-feeders mottled grey; the alpine ptarmigan white in winter, the red-grouse the colour of heather, and the black-grouse that of peaty earth, we must believe that these tints are of service to these birds and insects in preserving them from danger. Grouse... are known to suffer largely from birds of prey; and hawks are guided by eyesight to their prey... Hence I can see no reason to doubt that natural selection might be most effective in giving the proper colour to each kind of grouse...

ON

THE ORIGIN OF SPECIES

BY MEANS OF NATURAL SELECTION,

OR THE

PRESERVATION OF FAVOURED RACES IN THE STRUGGLE
FOR LIFE.

By CHARLES DARWIN, M.A.,
FELLOW OF THE ROYAL, GEOLOGICAL, LINNÆAN, ETC., SOCIETIES;
AUTHOR OF 'JOURNAL OF RESEARCHES DURING H. M. S. BEAGLE'S VOYAGE
ROUND THE WORLD.'

LONDON:
JOHN MURRAY, ALBEMARLE STREET.
1859.

The right of Translation is reserved.

The title page of Darwin's revolutionary book

This grouse would be hard to see against the heather. Darwin believed the bird's camouflage colouring was a product of natural selection that helped it hide from predators.

 Darwin's theory of natural selection was mocked by many people. They included the politician Benjamin Disraeli, who became British prime minister in 1868. The words he uttered on the subject, quoted here, have since become famous.

The question now placed before society is this, 'Is man an ape or an angel?' I am on the side of the angels.

GREGOR MENDEL
AND GENETICS

Charles Darwin (see pages 46-9) believed the offspring of any two creatures was a half-and-half mixture of its parents. Yet he also knew that if this inheritance pattern, known as blending, were true, natural selection would not work. An adult would pass on just half of a useful characteristic to its young, which in turn would pass on only a quarter. Darwin never solved this puzzle. But the Austrian monk Gregor Johann Mendel did.

In 1856, Mendel began to carry out plant-breeding experiments in the garden of the monastery in Brünn where he lived (see box). His aim was to show how different plant characteristics were passed from generation to generation. He decided the best way to do this would be to run several separate cross-breeding programmes, each tracking the path of just a single characteristic. The plants Mendel chose for his experiments were peas, and the seven characteristics he selected for study included height, pod colour and pea shape.

Mendel's first experiments involved cross-breeding tall and short pea plants. To his surprise, no plants of intermediate height were produced – they were all tall. This result immediately disproved the blending theory. However, when the new tall plants were cross-bred, one short plant was produced for

MENDEL'S PEA PLANT EXPERIMENT

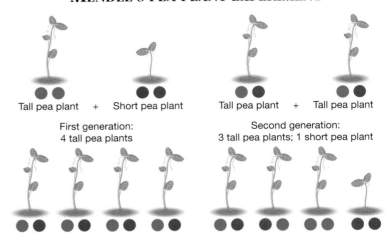

Tall pea plant + Short pea plant

First generation:
4 tall pea plants

Tall pea plant + Tall pea plant

Second generation:
3 tall pea plants; 1 short pea plant

This diagram of Mendel's pioneering experiment shows how the genes for tall and short were passed through two generations of plants.

about every three tall plants. This ratio was repeated in experiments for several other characteristics. Mendel continued this work for eight years, all the time painstakingly recording his results. Then he came to some important conclusions.

Inheritance, Mendel said, was governed by 'factors'. Each

GREGOR MENDEL (1822-84)

Gregor Johann Mendel (see picture below) was born in the Austrian town of Heinzendorf (now Hynčice in the Czech Republic). His father was a farmer, and as Mendel worked with him in the fields, he developed a great interest in plants. After studying at Olmütz University, Mendel became an Augustinian monk. From 1843, his home was a monastery in the town of Brünn that was already a centre of scientific research. It was there, between 1856 and 1863, that Mendel carried out his famous plant experiments. In 1868, Mendel became abbot of his monastery, and afterwards had little time for scientific study.

parent plant contributed a single factor for each characteristic to its offspring so that it had two in total. This combination determined which form of a characteristic it would inherit. If the two factors were the same, for example if both were for tallness, then the plant would inherit that trait. But if the two were different, then one factor, the dominant factor, would cancel out the other, the recessive factor. This theory explained the results of the pea plant experiments (see diagram).

In 1865, Mendel read his findings to the Brünn Natural History Society and published them in its journal (see document). Sadly no one took much notice. It was only after Mendel's death in 1884 that their significance was revealed. 'Factors' were what scientists now call genes and Mendel had founded the revolutionary science of genetics (see pages 56-7).

The following extract is taken from *Experiments in Plant Hybridisation* (cross-breeding), the paper Mendel read to the Brünn Natural History Society in 1865. In it, Mendel lists the seven pea plant characteristics that he examined.

AMENDING MENDEL

Hugo de Vries, Professor of Biology at Amsterdam University, rediscovered Mendel's work in 1900. Among the many scientists who read it shortly afterwards was Thomas Hunt Morgan (see picture right), Professor of Experimental Zoology at Columbia University, New York. He doubted Mendel's theories until 1908, when he began to breed fruit flies. The results of his own experiments proved Mendel was broadly

right. However, Morgan made one vital amendment to the monk's laws. Mendel had stated that each characteristic is passed by gene from parent to offspring independently. Morgan showed that groups of genes make up larger structures called chromosomes (see pages 56-7) and that genes on the same chromosome are often inherited together.

Mendel used this word to mean the endosperm, the food-material inside seeds that surrounds the growing embryo.

The characters which were selected for experiment relate:
1. To the difference in the form of the ripe seeds...
2. To the difference in the colour of the seed albumen...
3. To the difference in the colour of the seed-coat...
4. To the difference in the form of the ripe pods...
5. To the difference in the colour of the unripe pods...
6. To the difference in the position of the flowers...
7. To the difference in the length of the stem...

The first, handwritten page of the paper that Mendel read to the Brünn Natural History Society

LOUIS PASTEUR
AND GERM THEORY

Frenchman Louis Pasteur (1822-95) was both a chemist and a microbiologist – a scientist who studies organisms so small they can be seen only under a microscope. His major breakthrough was to discover that these tiny living creatures are responsible for many diseases. Armed with this knowledge, he also developed vaccines to prevent diseases spreading.

Pasteur became Professor of Chemistry at Lille University in 1854. The city of Lille was an important beer-brewing centre. One of the beer-making processes involved allowing sugar beet to ferment so that it turned into alcohol. But at the time of Pasteur's arrival, instead of forming alcohol, the sugar beet in many vats was turning sour. Pasteur was asked to find out why. He examined both the alcoholic and the sour liquids. In the alcoholic liquid, he saw specks of yeast, a type of microorganism. He worked out that they caused fermentation by breaking down sugar in the liquid to produce alcohol and carbon dioxide. No one had understood this process before. Pasteur found a different microorganism in the sour liquid. It was causing fermentation, too, but the product was lactic acid, not alcohol. Pasteur told the brewers to destroy all the liquid contaminated by these microorganisms. They did so, and the problem was solved.

Pasteur now wanted to discover where microorganisms came from. Were they in the air, as he believed, or spontaneously produced by rotting substances, as other scientists claimed? To find out, he did an experiment. First he put some sterilised fluid into two groups of swan-necked flasks (see picture). Then he snapped the necks off one group of flasks, allowing dust to enter. The fluid in these flasks was soon full of microorganisms, but the fluid in the other flasks remained clean. The 'germs', as Pasteur now called the microorganisms, were in the air.

Pasteur put his new understanding to practical use by inventing pasteurisation treatment – the heating of wine, milk and other liquids to high temperatures in order to kill harmful microorganisms. By this time he was convinced that such germs were the cause of many human and animal diseases, too. Soon he discovered the germ that caused the highly infectious disease anthrax, and in 1882 created an effective vaccine from a weakened form of this microorganism. In 1885, he

A replica of one of the swan-necked flasks that Louis Pasteur used in his germ theory experiments. It is filled with a nutrient-rich broth similar to the jelly Pasteur describes in the document on page 53.

Louis Pasteur. In 1857, Pasteur left Lille (see main text) for Paris. There he held several important posts, in particular Professor of Chemistry at the Sorbonne from 1867 to 1874.

produced a successful vaccine against rabies and dramatically used it to save the life of a young boy.

Pasteur became a public hero, and in 1888 a medical research institute named after him was opened in Paris. Today scientists at the Pasteur Institute still seek cures for diseases such as AIDS. If it were not for the discovery that many diseases are caused by germs, they would have little hope of success.

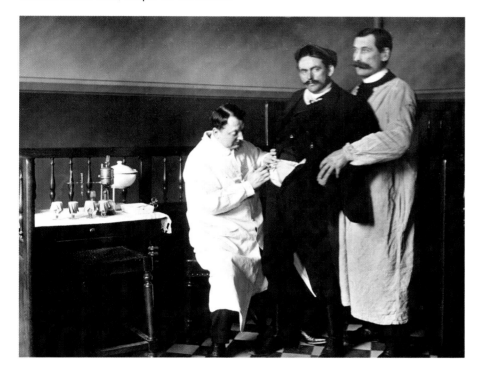

A man bitten by a rabid dog is given a vaccination to protect him from rabies at the Pasteur Institute in 1905.

SAVING THE SILKWORMS

In 1865, Pasteur was called away from Paris to southern France, where the silk industry was in crisis. The silkworms were dying in their thousands, so could not spin the shining threads on which the industry depended. The worm breeders had a name for the main disease causing all the damage – pebrine – but no idea how it was caused. Working with his usual thoroughness, Pasteur was able to identify the microorganism responsible. Before returning to Paris, Pasteur explained to the breeders how to make their industry disease-free and earned their gratitude.

Many 19th-century scientists disagreed with Pasteur's view that disease-causing microorganisms were present in the air. These scientists believed in 'spontaneous generation', the idea that diseased or decaying substances produced the microorganisms as they broke down. But Pasteur was unshakeable in his belief. In this extract from a lecture that he gave at the Sorbonne, he explains how he tried to grow microorganisms in a drop of water. Then he states why it did not work.

In all the immensity of creation, I have taken my drop of water and I have taken it full of rich jelly – that is, to use the language of science, full of elements most suited to the development of small beings. And I wait, I observe, I question it, I beg it to be so kind as to begin over again, just to please me, the primitive act of creation; it would be so fair a sight! But it is mute! It has been mute for years. Ah! That is because I have kept far from it, and still keep far from it, the only thing that it has not been given to man to produce. I have kept from it the germs that float in the air; I have kept from it life, for life is a germ and a germ is life. Never will the belief in spontaneous generation arise from the mortal blow that this simple experiment has given it.

ALEXANDER FLEMING
AND PENICILLIN

This petri dish demonstrates the action of penicillin. The mould is inhibiting the growth of some of the bacteria streaks on the jelly, just as it did in the original dish discovered in 1928.

other scientists was low, and penicillin research gradually petered out.

In 1939 two other scientists, Australian Howard Florey and German-born Ernst Chain (see box page 55), returned to the subject. As the Second World War began, the British government encouraged their research, hoping it might lead to new battlefield treatments. By 1941, the scientists had produced a pure form of the penicillin that worked as a drug. Wartime Britain, however, could not afford to make it, so Florey turned to the USA for help.

Louis Pasteur's germ theory (see pages 52-3) was finally accepted, giving scientists a far better understanding of disease. However, many types of germ remained a threat to human and animal life. The Scottish bacteriologist Alexander Fleming made a discovery that was to provide doctors with a powerful new way of overcoming them.

In 1928, Fleming was appointed Professor of Bacteriology at St Mary's Hospital in London, and in the same year he made his most significant discovery. As part of his research work into the treatment of influenza, Fleming had to grow bacteria on jelly in special glass dishes. One morning he noticed that something had destroyed the bacteria on certain areas of jelly in an uncovered dish. Tests showed that this was a substance produced by a mould whose Latin name was *Penicillium notatum*. Fleming called it 'penicillin'.

Fleming knew that penicillin was effective against staphylococcus, the type of bacteria on the dish. Now he began experiments to discover which other types of bacteria it would destroy. The results were encouraging, but Fleming was unable to isolate the substance that killed the bacteria. Interest among

ALEXANDER FLEMING (1881-1955)

Alexander Fleming was born in Lochfield, Scotland. When he was 16, he left school and took a job in London as a shipping clerk. But he wanted to pursue a medical career, so began a course of study at St Mary's Hospital Medical School in the west of the capital. He graduated in 1908, but stayed at the school, leaving only temporarily to serve in the First World War. Once back, he devoted himself to the study of bacteria and their control. In 1922, he discovered lysozyme, an enzyme present in tears, saliva and other body fluids that is an effective antiseptic. Six years later, while Professor of Bacteriology at St Mary's (see main text), Fleming made the far greater discovery for which he is now famous.

Fleming in his laboratory at St Mary's Hospital, Paddington, in 1909

Mass-production of penicillin began there in 1943 and a few years later this new antibiotic became widely available.

The work of Fleming, Florey and Chain ushered in a new era of medicine. Soon doctors were using penicillin and other antibiotics to treat many diseases. But now, in the 21st century, they are more wary. Overuse of antibiotics has enabled some bacteria to develop resistance to them. So today the search is on for new types of these life-saving drugs.

FLOREY AND CHAIN

Howard Florey and Ernst Chain (see main text) came from opposite sides of the world. Florey was born in Adelaide, Australia in 1898 and studied there before winning a scholarship to Oxford University in England. He returned to Oxford in 1935 as head of the Sir William Dunn School of Pathology. Chain was born in Berlin, Germany, in 1906. But in 1933, when the Nazis came to power, he left for Britain joining Oxford University in the same year as Florey. Ten years later, Florey, Chain and Fleming shared the Nobel Prize for their work on penicillin.

The Nobel medal awarded to Florey, Chain and Fleming in 1945

By the mid-1940s, doctors were able to prescribe penicillin. But it was still a new drug, so they needed guidance on how to use it. To help them, one publisher brought out a book called *Penicillin – Its Practical Application* (1946). The book was edited by Alexander Fleming, who also wrote the introduction. In this extract from the introduction, he describes one of the methods that he used to test penicillin.

Petri dishes are shallow, round glass dishes with lids. They are used in laboratories for growing microorganisms.

Microbes here means 'bacteria'.

Incubation is the period during which the bacteria were left to grow.

Gutter here means 'groove'.

With a knife a gutter was cut out of the agar in a Petri dish and was filled with agar containing... the mould. Different microbes were then planted in streaks from the edge of the plate to the gutter. After incubation some of these were found to grow right up to the gutter while others were inhibited for a considerable distance. This showed that the mould... had a powerful inhibitory action on some bacteria but not on others, and that among the sensitive bacteria were some of the commonest agents of infection in man.

Agar is a type of jelly on which bacteria are grown.

Inhibited here means 'held back' or 'prevented from growing'.

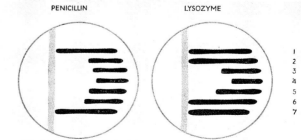

FIG. 7.—Gutter method of demonstrating selective inhibition

PENICILLIN	LYSOZYME
1. *Esch. coli*	1. *Esch. coli*
2. Staphylococcus	2. Staphylococcus
3. Haemolytic streptococcus	3. Sarcina
4. Gonococcus	4. *M. lysodiekticus*
5. *B. diphtheriae*	5. *B. subtilis*
6. *B. anthracis*	6. Streptococcus
7. *B. typhosus*	7. *B. typhosus*

This diagram from the book shows how the penicillin in the gutter (left-hand image) stopped bacteria of types 2 to 6 from growing.

CRICK AND WATSON AND DNA

By the 1950s, thanks to Mendel, Morgan (see pages 50-51) and others, scientists had a basic understanding of the role played by genes and chromosomes in passing characteristics from one generation to the next. They also knew the chemical composition of these vital pieces of matter. But their structure remained a mystery until Anglo-American duo Francis Crick and James Watson made a major breakthrough.

In 1869, Swiss biochemist Friedrich Miescher had discovered a substance that seemed to occur in all cell nuclei. He called it nuclein, but it later became known as nucleic acid. There were at least two types. Other scientists then began to work out what one particular type of nucleic acid contained. This turned out to be four chemical compounds of a kind known as bases, another type of chemical called a phosphate, and a sugar

called deoxyribose. To distinguish it from other types, the acid was given its own special name of deoxyribonucleic acid – DNA.

Next, scientists turned their attention to two more puzzles. What was the exact location of DNA in cells, and what was its purpose? Both questions were answered by the 1940s. German chemist Robert Feulgen showed that DNA was found in chromosomes, and in the genes they contained. Canadian

FRANCIS CRICK (1916-) AND JAMES WATSON (1928-)

Francis Harry Compton Crick was born in Northampton, England, and James Dewey Watson in Chicago, USA. Crick studied physics at University College, London and Cambridge University. During the Second World War, he was employed by the Admiralty to design new types of mine. Watson studied at the University of Chicago and the University of Indiana before moving to Copenhagen University and then to the Medical Research Council Unit at Cambridge. It was here that he met Crick. After their famous collaboration Crick remained at Cambridge until 1977, when he joined the Salk Institute in San Diego, USA. There he devoted himself to studying the brain and the nature of consciousness. In 1953, after the research on DNA structure was complete, Watson returned to the USA, eventually taking up a post at the Cold Springs Harbor Laboratory, New York. From 1988 to 1992, Watson took a leading role in the ground-breaking Human Genome Project (see page 58).

James Watson (left) and Francis Crick with their model of part of a DNA molecule in 1953

A DNA STRAND

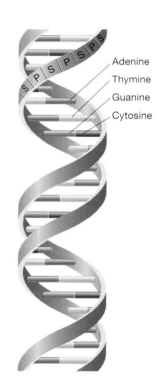

Adenine
Thymine
Guanine
Cytosine

The four bases in DNA – adenine, thymine, guanine and cytosine – make up the 'rungs' of its twisted ladder shape. The sides of the ladder consist of alternating strips of phosphate (P) and deoxyribose sugar (S).

chemist Oswald Avery and his team showed that DNA from one kind of bacteria altered the nature of a second kind in which it was placed. In other words, DNA carried the hereditary material that produced the characteristics of living things.

Many scientists were now eager to learn more about DNA. They included Englishman Francis Crick and American James Watson (see box page 56). By the 1950s, both were working in the Cavendish Laboratory at Cambridge University. Watson persuaded Crick to join him in the task of discovering the structure of DNA and the two men set to work. However, many other scientists had already carried out important research on the subject. Crick and Watson relied on X-ray diffraction pictures of DNA provided by Maurice Wilkins and Rosalind Franklin (see box). Nevertheless, it was their thought-processes that finally revealed the shape of the DNA molecule. This was a graceful double helix that looked rather like a twisted ladder.

The significance of Crick and Watson's work was enormous. Knowledge of DNA's structure enabled scientists to work out exactly how genetic messages are passed between generations, leading to a deeper understanding of heredity and evolution. It also led to the development of genetic engineering – the alteration of genes to produce improved types of plants, animals, bacteria, vaccines and more.

ROSALIND FRANKLIN

The role of Rosalind Franklin (see picture below) in discovering the structure of DNA is much debated. Franklin was born in London and educated at Cambridge University. She joined King's College, part of London University, in 1951. While there, she was asked to help with the DNA project. Unfortunately, there was a personality clash between Franklin and Maurice Wilkins, who was also working on the DNA puzzle at King's. In addition, Franklin felt that her scientific contribution was continually underestimated because she was a woman. Most historians of science now accept that her X-ray diffraction photographs provided Crick and Watson with some of the most important evidence for their discovery. But there is disagreement about whether she would ever have worked out the structure of DNA herself.

In 1988, Francis Crick published *What Mad Pursuit*, a book that records his part in the DNA story and describes his later research. In this extract, he spells out what made the DNA project just like any other piece of scientific research – and what made it different.

I think what needs to be emphasized about the discovery of the double helix is that the path to it was, scientifically speaking, fairly commonplace. What was important was not the way it was discovered but the object discovered–the structure of DNA itself. You can see this by comparing it with almost any other scientific discovery. Misleading data, false ideas, problems of personal interrelationships occur in much if not all scientific work. Consider, for example, the discovery of the basic structure of collagen, the major protein of tendons, cartilage, and other tissues. The basic fiber of collagen is made of three long chains wound around one another. Its discovery had all the elements that surrounded the discovery of the double helix... Yet nobody has written even one book about the race for the triple helix. This is surely because, in a very real sense, collagen is not as important a molecule as DNA... It is the molecule that has the glamour, not the scientists.

CONCLUSION

The achievements of scientists over the last two centuries have been mighty. Their practical skills and theoretical brilliance have led them to discover everything from tiny subatomic particles and microorganisms to the powerful mechanisms that drive the Earth and the stars.

Such results did not come without effort. Many of the scientists featured in this book, for example Marie Curie, spent years carrying out painstaking experiments. Others, such as Charles Lyell, had to scour the Earth for evidence to support their theories. Albert Einstein sometimes became ill as he wrestled with complex mathematical problems. Despite all this hard work, setbacks were common. When breakthroughs were made, they were often mocked, as in the case of Alfred Wegener. Occasionally they caused public outrage, as Charles Darwin discovered.

Of course the problems of research did not, in the end, deter the great scientists of the last 200 years. Sheer curiosity, as well as the support of individual fellow scientists, and sometimes of the worldwide scientific community, spurred them on. Almost all record the great pleasure that their discoveries gave them – Louis Pasteur said 'the reward is one of the keenest joys of which the

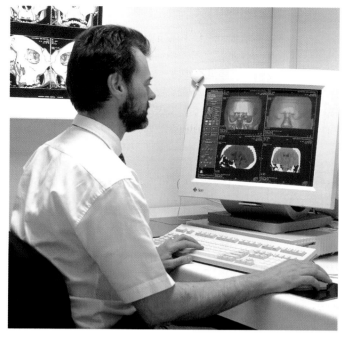

Pure science has led to many beneficial technological advances. For example, the discovery of X-rays made possible the development of CT (computerised tomography) scanners like this one. They use X-rays to build up three-dimensional pictures of the body as an aid to medical diagnosis.

human soul is capable'. Often this was all the greater if the discovery had taken a long time to make. Public recognition in the form of a Nobel Prize was another boost, even if awarded long after their work was complete.

In the 21st century, the role of the scientist is likely to be greater than ever. Discoveries in the

In 2000, scientists completed their map of the human genome, that is all the genes that humans contain. Now they know exactly where each one is on our chromosomes, and are learning what each one does. The DNA of every person's genes is different producing a 'fingerprint' – a unique pattern of black bands (seen here).

fields of genetic engineering, artificial intelligence and communications will almost certainly transform the future, just as discoveries relating to electricity, atomic structure, evolution and more transformed the past. But how this will happen, and what else may happen, is completely impossible to predict.

Recent centuries have provided ample proof that scientific discoveries can be put to destructive as well as to good use. Albert Einstein's realisation that energy is locked up in matter eventually led to the creation of the atomic bomb and its dropping on Japan (see page 36). In many people's view, this is the most obvious example of science gone astray. Robert Oppenheimer was director of the laboratory in Los Alamos, USA, that made the bomb. Here he explains how tempting it is to continue research without thinking clearly about the consequences.

When an atomic bomb is dropped, a distinctive mushroom cloud appears over the site. This cloud appeared after a nuclear test in Nevada, USA.

...it is my judgement in these things that when you see something that is technically sweet you go ahead and do it and you argue about what to do about it only after you have had your technical success. That is the way it was with the atomic bomb. I do not think anybody opposed making it; there were some debates about what to do with it after it was made.

Louis Pasteur also recognised that science could be used for harmful purposes. But he knew from his own achievements that it could also be a force for good. In a speech he wrote for his 70th birthday celebrations, he stated his belief that the good would triumph.

Gentlemen, you bring me the greatest happiness that can be experienced by a man whose invincible belief is that science and peace will triumph over ignorance and war...Have faith that in the long run...the future will belong not to the conquerors but to the saviors of mankind.

GLOSSARY

absolute magnitude the brightness of a star at a standard distance of 32.6 light years from the Earth.

alpha particle a type of particle made up of two protons and two neutrons. Alpha particles are emitted from atomic nuclei during some types of radioactive decay.

antibiotic any of a class of drugs used to reduce or stop the growth of disease-causing bacteria.

apparent magnitude the brightness of a star as it appears from Earth.

artificial intelligence (AI) advanced computer systems and programs that try to mimic human thought processes.

astrophysics the branch of astronomy that deals with the chemical composition, origin and evolution of stars and other heavenly bodies.

atom the smallest particle of an element that is characteristic of that element and that can take part in chemical reactions.

atomic bomb a type of bomb whose explosion is caused by splitting atomic nuclei.

atomic number the number of protons in the nucleus of an atom. The atoms of each element have a different atomic number.

atomic weight the old name for relative atomic mass.

bacteriologist a scientist who studies the nature and behaviour of bacteria.

base any of a group of chemical compounds that, when combined with an acid, form another type of chemical, called a salt, and water.

beta particle an electron, or its positively charged opposite, a positron, emitted from atomic nuclei during some types of radioactive decay.

Big Bang The vast explosion that most scientists believe took place between about 15 and 17 billion years ago which hurled outwards all the matter in the universe.

biochemist a scientist who studies the chemical composition of living things.

black hole any area in space believed to be the collapsed remains of a star with a mass more than three times the Sun's. The superdense mass creates a powerful gravitational field from which no light is able to emerge.

cathode the negatively charged part of electronic devices such as valves and vacuum tubes that produces a stream of electrons.

chromosomes 23 pairs of tiny, rod-shaped structures that occur in the nucleus of every human cell. They are made up of thousands of genes, which transmit hereditary characteristics from parents to children.

compound a substance made up of two or more elements whose atoms are joined together chemically.

condensed changed from a gas to a liquid.

continental drift the gradual movement of the continents over the surface of the Earth.

convection current the constant movement of material below the Earth's surface. Convection currents drive the movement of the Earth's plates.

cosmology the study of the formation, structure and evolution of the universe.

electrolysis the passing of electricity through a solution or molten solid in order to produce chemical changes.

electromagnetic spectrum the full range of electromagnetic waves, from long-wavelength radio waves to short-wavelength gamma rays.

electron a type of negatively charged particle that is present in all atoms and that orbits the nucleus.

element a substance that contains only one type of atom, which all have the same number of protons in their nuclei.

enzyme any of a group of proteins that occur in cells and speed up the biochemical reactions that take place there.

erosion the gradual wearing away of the Earth's surface by wind, rivers, seas and ice.

fold a bend in rock that is caused by sideways pressure in the Earth's crust.

fossil the remains of an ancient plant or animal that have been preserved in rock.

galaxy a large group of stars that is held together by gravity and whose members all orbit around a central point. There are four main galaxy types: spiral, barred spiral, elliptical and irregular.

gene one of many thousands of tiny structures that make up chromosomes and that pass characteristics between generations. Different genes carry different characteristics,

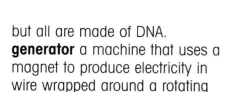

but all are made of DNA.

generator a machine that uses a magnet to produce electricity in wire wrapped around a rotating iron core.

genetics the study of the structure and behaviour of genes.

gravitational field the area around a body in which the effects of its gravity are experienced.

gravity the force of attraction between objects that have mass.

insulin a hormone produced in an organ of the body called the pancreas, which lies behind the stomach. A shortage of insulin, which controls the amount of sugar in the blood, leads to diabetes.

luminosity the amount of light energy emitted by a star.

magma hot, molten rock.

magnetic field the area around a magnet in which the effects of its magnetism are experienced.

mass the amount of matter in an object.

matter any substance that occupies space. Most matter exists as a solid, a liquid or a gas.

mixture a substance made up of two or more elements whose atoms are simply mixed together, not chemically combined.

molecule the smallest particle of a substance that has its chemical properties. Molecules of elements are made up of two or more chemically joined atoms of the same type. Molecules of compounds are made up of two or more chemically joined atoms of different elements.

motor effect the movement produced in an electric current-carrying wire when it is placed in a magnetic field.

neutron a subatomic particle that carries no electric charge, positive or negative. The nuclei of all atoms except hydrogen atoms contain neutrons.

neutron star a dense, collapsed star that has a mass 1.4 to 3 times that of the Sun and that is made up entirely of neutrons.

nucleus the positively charged central region of an atom around which negatively charged electrons orbit. Most atomic nuclei are made up of protons and neutrons, but hydrogen nuclei contain only a single proton.

oceanographer a scientist who studies the origin, structure and development of the oceans and ocean floors.

optics the branch of science that involves the study of vision and light.

PhD The abbreviation for Doctor of Philosophy. This title is granted to people who have gained a doctorate by carrying out original research.

photographic plate a sheet of metal or glass coated with light-sensitive chemicals.

proton a subatomic particle with a positive electric charge. The nuclei of all atoms contain protons, but the number varies. The number of protons in a particular element is known as its atomic number.

radiation the emission of energy from a body in the form of waves or particles. Heat, light, sound and radioactivity are all forms of radiation.

radioactive decay the emission of particles or rays from the atomic nuclei of elements such as radium as they spontaneously disintegrate.

radio telescope a telescope that is designed to detect radio waves

rather than visible light from sources in space.

red supergiant any of a class of stars that have a higher luminosity than ordinary red giants and are also larger.

relative atomic mass the number of times the mass of an atom of an element is greater or smaller than the mass of one-twelfth of a carbon-12 atom.

rift a huge crack in the Earth's surface caused by two plates moving apart.

sedimentation the settling or depositing of rock, soil and other particles (sediment) produced during erosion.

solubility the ability of a substance to dissolve in water or other liquid.

sonar An apparatus that detects the position of underwater objects. It does so by sending out sound waves that bounce back when they hit an object, revealing the object's position.

spectroscopy the science of examining light spectra.

spectrum (plural **spectra**) the full range of colours produced when white light is dispersed, for example by passing it through a prism.

strata (singular **stratum**) layers, especially layers of rock of different ages and types.

subatomic particle any of the particles that make up, and are smaller than, atoms. Subatomic particles include electrons, protons and neutrons.

supernova a star with a mass at least eight times the Sun's that collapses once it has burned up all its fuel, then explodes. The explosion causes it to become intensely bright and extremely hot – up to 10,000 million°C.

INDEX